Redesigning the American Lawn

Redesigning the American Lawn

A Search for Environmental Harmony

F. Herbert Bormann

Diana Balmori

Gordon T. Geballe

Lisa Vernegaard, Editor-Researcher

Yale University Press

New Haven and London

Set in New Century Schoolbook type by The Composing Room of Michigan, Grand Rapids, Michigan and printed in the United States of America by Vail-Ballou Press, Binghamton, New York.

Library of Congress Cataloging-in-Publication Data
Bormann, F. Herbert, 1922–
The American lawn reconsidered : lessons from ecology and art /
F. Herbert Bormann, Diana Balmori, Gordon T. Geballe ; Lisa
Vernegaard, editor-researcher
p.   cm.
Includes bibliographical references (p.   ) and index.
ISBN 0-300-05401-7
1. Lawns—United States.   2. Lawn ecology—United States.
I. Balmori, Diana.   II. Geballe, Gordon T., 1947–
III. Vernegaard, Lisa.   IV. Title.
SB433.B64   1993
635.9′647′0973—dc20   92-42918 CIP

A catalogue record for this book is available from the British Library.

The paper in this book meets the guidelines for permanence and durability of the Committee on Production Guidelines for Book Longevity of the Council on Library Resources.

10   9   8   7   6   5   4   3   2   1

Contents

The Genesis of This Book

This book represents a collaboration between the faculty and students of the School of Forestry and Environmental Studies and the School of Art and Architecture at Yale University. "The American Lawn," a graduate seminar, was offered at Yale University during the spring semester of 1991. The following graduate student authors contributed to this book:

Jon H. Connolly

Jennifer Greenfeld

Anne H. Harper

Lee Ann Jackson

William L. Kenny

Barbara Milton

John Petersen

Susan L. Pultz

Chris Rodstrom

Lisa Vernegaard

Jennie M. Wood

As environmentalists and educators, the authors of this volume have been concerned with environmental education for many years. Environmental problems are most often complex and difficult to understand. Teaching environmental studies in a coherent and integrative way is no easy task. In the fall of 1990, the first author proposed that the American lawn would be an excellent vehicle for teaching environmental principles. Soon joined by his two colleagues, the team designed a course for the spring of 1991, with the American lawn as its central theme. The goal of the course was truly unique—to compile a book-length manuscript on the topic, the American lawn: is it an environmental anachronism? With the assistance of Dr. Joseph Miller, librarian and historian, the eleven graduate students examined and modified the chapter outlines proposed by the three author-instructors. Student teams were assigned to write each chapter. After reading the literature and interviewing lawn experts and knowledgeable faculty members, the teams wrote their first drafts. These were distributed to all participants and subjected to public and private criticism. The drafts were rewritten and the process repeated. Final drafts were subjected to another round of criticism. The present chapters are the work of the students, modified, substantially rewritten, and added to by the authors.

All proceeds from this book will be used to provide fellowships for students interested in ecology at the School of Forestry and Environmental Studies and at the School of Architecture at Yale University.

Acknowledgments

This book is a collaboration of many people and organizations. The authors would like to thank all who gave of their time, answered our many queries, and patiently assisted us.

This book could not have come into being without the persistence and assistance of the staff at Yale University Press. Jean E. Thomson Black, science editor, encouraged us from the beginning.

Anyone who has worked at Yale University knows of the invaluable resource provided by the library system and its staff. Joseph Miller of the Henry S. Graves Memorial Library at the School of Forestry and Environmental Studies provided expert guidance and readily responded to requests for assistance.

All the authors called upon the assistance of those who work with them. Audrey Anderson assisted with research. Carol Ziegler helped to produce countless drafts of the text.

Many individuals gave freely of their time. Our thanks go to Steven Beissinger, Christine Bormann, Dan Botkin, William Dest, Larry Forcier, Frank Golley, Morgan Grove, Jean-Marie Hartman, Gene Likens, Darrell Morrison, Carleton Ray, William Smith, Bill Welsh, and George Woodwell.

The book's illustrations were produced by several talented artists and photographers. Those who worked closely with us include Lauren Brown, Karen Bussolini, Tony Casper, and Alex Taylor.

We contacted several organizations throughout the country while conducting our research. Special thanks go to Eliot C. and Beverly C. Roberts of the Lawn Institute.

This book could not have been produced without the backing of two philanthropic organizations. The Mary Flagler Cary Trust with Ned Ames as executive director provided the financial backing that supported the development of the book; Ned's quick and enthusiastic response encouraged us and confirmed the usefulness of this project. The Andrew W. Mellon Foundation and William Robertson stood ready to assist, permitting us to move ahead with minimum anxiety.

Prologue

Americans' attachment to the lawn is a long and fond one. A lawn is a gathering place for family, friends, and neighbors, a place where we engage in our favorite activities. In cities, it is a place of verdure, a refuge from crowds, traffic, and noise. The green blades feel good to the touch; the cut grass freshens the smell of the air. No other nation, except perhaps England, holds the lawn in such reverence. In passing through suburban neighborhoods where one landscaped lawn follows another, we can vividly see the pride Americans take in their lawns.

Our long love affair with the lawn has had its rough spots, but none more critical than its recent implication as another factor in the deterioration of the environment of the earth. What possible connection can there be between the lawn and the earth's biosphere? It is the purpose of this book to explore the numerous connections, to point out the many ways that we as lawn owners through our lawn management practices diminish in small but collectively significant ways local, regional, and global environments, and finally to suggest ways by which we can enjoy the many virtues of the lawn while reducing our impact on nature.

For many of us, the realization that our actions may be contributing to the deterioration of the planet is recent. About thirty years ago, signs of a steady decline in environmental quality became visible: polluted streams and rivers, smog-shrouded cities, and urban decay were everywhere. Scientists began to document more subtle effects: food chains contaminated with pesticides and air masses polluted with an extraordinary variety of wastes from the enormous engine that throbs at the heart of our society.

What seemed a patchwork of environmental problems, each specific to a particular region or country, began to coalesce into global phenomena: global warming caused by the accumulation of greenhouse gases in the atmosphere; large regions of the earth affected by acid rain with the potential for serious damage to streams, lakes, and forests; and chlorofluorocarbon pollutants that thin the stratospheric ozone layer with a resultant increase in biologically destructive ultraviolet light reaching the surface of the earth.

All this is beginning to reveal to us a global environmental deficit, the unanticipated consequence of humanity's alteration of the earth's atmosphere, water, soil, flora, fauna, and ecological systems. We are beginning to see that human activities may be disrupting the very life systems on which we all depend.

What place does the American lawn occupy in this scenario? Whether suburban backyard or city park, a gracious expanse or a tiny strip of green, the lawn is part of the earth's surface. It is the most commonplace landscape and the one most familiar to us. For the homeowner, the lawn is also our piece of the biosphere, and through it we communicate our concern about the environment of the earth, our greater yard.

To see the effects of ecological deterioration on our own lawns or in our own backyards is to have abstract ecological problems brought home. Issues of geochemical and hydrologic cycles, soil formation and species diversity seem abstractly important but beyond our reach. Ecological disasters that can contaminate the groundwater of whole regions of the Midwest or wipe out the tropical forests of a whole continent seem almost beyond our comprehension, let alone our ability to prevent. Yet, ecologically destructive practices that threaten the global environment are present in smaller landscapes, including our lawns. Understanding the dynamics of lawn ecology may bring to a human scale the meaning of ecological sustainability. By the same token, understanding this small ecological system may impel us to constructive action. We can begin to see that our weekend puttering on the lawn can mean caring for the planet. Fostering the development of an ethic of "environmental awareness" and exploring ways of implementing that ethic on the small piece of the environment entrusted to our care are the dual purposes of this volume.

To understand how our actions in managing the lawn might affect the biosphere, it is necessary to understand how the naturally occurring ecological systems—ecosystems—that the lawn replaces function. Forests, prairies, or other naturally occurring ecosystems are composed of large numbers of plants, animals, and microorganisms. Ecosystems function both above and below ground; they change with the seasons; but most of all they are systems powered by the sun (see figure 1).

Using solar energy, ecosystems carry out an extraordinary array of processes. They store and recycle nutrients such as carbon and nitrogen, holding them within the system instead of releasing them to interconnected streams, lakes, and groundwater. They provide a home for the many species of organisms that live within their boundaries. Naturally occurring ecosystems make major contributions to the stability of the earth through their maintenance of air, water, and soil conditions favorable to human society everywhere on the planet. All of this is done using only the energy of the sun.

Human civilizations can affect even the most remote ecosystems through air and water pollution, and many naturally occurring ecosystems have been drastically changed by urbanization, agriculture, and forestry. Many such changes are essential if our societies are to exist in their present form. Yet perhaps we are going too far in our manipulation of nature, so far as to be damaging the very life-support systems upon which we depend. Pre-

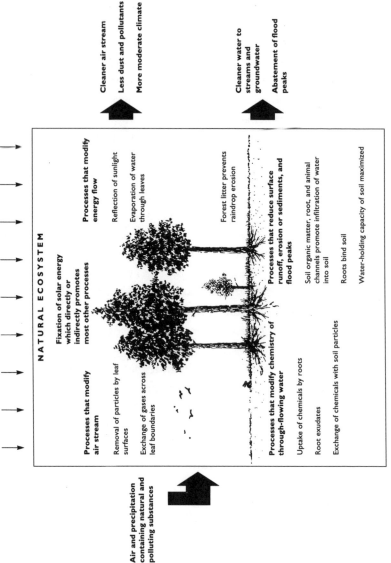

Figure 1. Using solar energy, naturally occurring ecosystems such as forests, prairies, and fields modify air and water in ways beneficial to human society. ©1985 American Institute of Biological Sciences.

dictions of environmental catastrophes within our children's lifetimes are no longer considered absurd. The 1992 Earth Summit at Rio de Janeiro, Brazil, reflected the concerns of people throughout the world for the health of the planet. It is in this light that we wish to examine the American lawn.

The American lawn is a human-modified ecosystem, and many questions arise concerning its ecological function. Does it, like naturally occurring ecosystems, contribute to global ecological stability? Or is it part of the problem? Alternately, in designing and managing our lawns, have we departed too far from nature's plan? If so, can we modify our lawn care practices to mitigate damage to the environment?

We have all grown up with the American lawn, but we must reevaluate our attachment to the lawn in light of continuing evidence that human activity is disrupting the biosphere. To replace the eighteenth-century notions of the beautiful lawn landscape is to shape a new aesthetic to go with our new ecological ethic. These new visions of our landscapes, ecologically sound and aesthetically pleasing, might also guide the way we build our cities and communities, and, in fact, the way we conduct our lives.

Chapter 1
Love of the Lawn

Any Street, USA

The lawn holds an important place in the American view of an ideal life. Over vast areas of small towns and suburbs a spreading green carpet forms the background for living. The well-kept lawn is not only beautiful in itself, it also provides the setting for the house. These are the homes where people feel in harmony with their well-tended plots of land. Coming home after work one might loosen a tie or kick off one's shoes and feel proud and relaxed without knowing why. In the spring new blades of grass emerge, creating a carpet of green that provides a perfect backdrop for the beauty of crocus, daffodil, azalea, and rhododendron. In summer the world is lush with vegetation and the air perfumed by freshly cut grass. The homeowner is filled with a sense of well-being. There are many variations depending on season, location, the resources and energy of the residents, but the ideal persists. The well-kept front lawns roll down the street providing open space and beautiful vistas. In this ideal, the grass sward is as pure as possible, mowed two inches high, and free from dandelions and other insidious intruders (see figures 2 and 3).

Behind these similar front lawns lie more varied backyards; some contain children's play equipment; others have patios, picnic tables, and barbecue grills; still others have gardens of vegetables or flowers. Here and there is an occasional pool. Unfinished tasks lie about: a pile of wood half stacked from last

Figure 2. A suburban lawn. Photograph by Lisa Vernegaard.

Figure 3. Connecticut Street, Litchfield, Connecticut. In suburbs and small towns throughout America, front lawns run together without interruption, giving a neighborhood a sense of unity and providing a source of community pride. Photograph by Diana Balmori.

winter, a building project, perhaps lumber for a tree house lies covered with plastic. Clothes hang on clotheslines. Under all of these activities the lawn rolls on, with bare spots marking the heavily used areas.

Whereas the front lawn is a bit like the parlor, the back lawn is more like the kitchen. The owner may be willing to share the front lawn for community functions, but privacy is what matters most in the backyard. You can step out on the back lawn, take a deep breath, and feel the sun on your face. It is just for you. In the front yard, at any moment a social obligation may arise, but in the backyard, you can set up a hammock, lie down, close your eyes, and relax. The backyard is yours alone.

Why We Love the Lawn

Anyone who has ever played croquet or soccer on it or laid a blanket over it to listen to a summer concert or watch a fireworks display has no need to ask why Americans love the lawn. But if we look below the surface, our love of the lawn is more complicated. It involves aesthetics, economics, psychology, and, especially, history.

The lawn expresses a familiar aesthetic: its green expanse provides the framework for flower beds and shrubs, majestic trees, and our homes. Its predictable horizontality presents a counterpoint to the verticality of trees, homes, and flowers, while at the same time enhancing details of their surfaces. Shrubs and trees bordering the edge in turn complement the flat landscape they enclose.

The grass sward leads the eye of the viewer toward the horizon and into distant areas. Landscape features that begin in the foreground may be followed toward a distant and nebulous endpoint on the horizon. The lawn has been repeated over and over again, millions of times, and yet it continues to fill us with delight and appreciation (figure 4).

Figure 4. From the air, we have a dramatic view of the dominance of lawns in suburbia. Our views of nature, the games we play, and our weekend chores are heavily influenced by this choice of landscape. Photograph by Barrie Rokeach/Image Bank.

Practical advantages should also be given their due. Some of the lawn's most beneficial functions are safety and health related. In some areas grass can serve as a firebreak, helping to keep wildfires at bay. Even before scientists discovered that turf can trap some pollutants and pollens that cause allergies, Walt Whitman called grass "the handkerchief of the Lord."[1] Lawns have great recreational value. They are home to baseball, croquet, touch football, badminton, and tag. Grass wears well and provides a cushioning effect that reduces injuries and makes walking, running, and jumping more comfortable (figure 5).

The clean, cool, natural greenness of a beautiful lawn provides a pleasant environment. Its soft surface reduces glare and noise and conspires to dim, just a bit, the hustle and drive of our generally demanding society. On any northern college campus, hundreds of students spread-eagled on every available patch of lawn bear witness to the comfort the lawn offers on the first warm days of spring.

Perhaps our love of the lawn contains elements that are both evolutionary and psychological. John Falk, who has been studying human preference for grassed landscapes for the past fifteen years, sampled a segment of the U.S. population representing diverse age groups and nationalities to get a sense of their landscape preferences. Overwhelmingly, people favored short grass and scattered trees. Children especially preferred short grass.

Since humans evolved in the grassy, tree-sprinkled savannas of Africa, Falk suggests that our modern preference for lawns and trees is an innate expression of our origins.[2] The evidence for adapted predisposition is circumstantial at best, but Falk maintains that the connection is not unreasonable: "If we buy the notion of genetic vestige," he says, "then humans will find lawns a habitable, safe, and potentially supporting environment." Why does he think people love lawns? Falk laughs and replies, "Because we can't help ourselves."

Some psychologists agree that humans have an innate prefer-

Figure 5. Lawns are safe havens for kids of all ages. Ballgames, tag, picnics, and naps are often our first experiences with nature. The songs of birds and the smell of newly mowed grass are two of the most relaxing sensations suburbanites enjoy. Photograph by Karen Bussolini.

ence for open spaces. Open spaces provide "legibility," an environment that is clear and easily understood, where people are more likely to acquire information and less likely to get lost.[3]

The lawn even has its political overtones. Thomas Jefferson laid the groundwork for lawn ownership by advocating an order where "as few as possible shall be without a small portion of land."[4] Our second president, John Adams, proposed "the only possible way then of preserving the public virtue is to make the acquisition of land easy to every member of society: to make the dividing of land into small quantities that the multitudes may be possessed of landed estates."[5] Adams and Jefferson may not have been thinking of the lawn when they advocated universal ownership of small plots as the foundation for American democracy, but such ownership is indeed the basis for our millions of lawns.

Unquestionably economics plays a major role in our "love" of the lawn. A home is the cornerstone of many people's net worth—their primary asset. Great efforts are expended to maintain the home's value; since it is believed that landscaping can add up to 15 percent to a home's worth, lawns contribute to resale value.[6]

History of the Lawn

Given the reasons just cited, it would seem—particularly if genetic preferences have any validity—that all cultures throughout history would have made lawns central to their public and domestic spaces. Yet this is not the case. Lawns arose primarily in Western civilization, and mainly in France and England, and it is the English part of its history that is particularly important to our love of the lawn.

In terms of human history, the lawn is not an old tradition. Its popularity began in eighteenth-century Europe, but its antecedents are deeply embedded in humankind's struggle to understand and control nature.

In medieval times, gardens were a form of environmental control: surrounded by walls, they provided a psychological sanctuary for human activity. Practical gardens provided a cultivated area where food and herbal medicines were grown, while in pleasure gardens, fruits and flowers invited lounging, dancing, and romance (plate 1).[7] Gardeners used the limited space in the walled gardens carefully, espaliering, dwarfing, and raising planters over bare rock at times. To save space, they often planted crops such as fruit trees in staggered rows, with the trees closely trimmed. The tight geometric composition of productive gardens behind walls may have been the forerunner of the geometric layout of agricultural fields and of the seventeenth-century French aristocratic gardens. As political power in France shifted from the individual walled fiefdoms to a centralized monarchy, the walls of these aristocratic gardens slowly came down (plate 2).

In the open French formal gardens of the seventeenth century, gardeners experimented with raised beds of grass shaped in ornate patterns. Behind these precise patterns lay a belief in a nature that worked with clocklike precision, and the image of a clock often served as a metaphor for the hidden workings of the universe. Visitors strolled on walks of gravel woven through elevated grassy beds shaped in the precise yet graceful curves and arabesques of embroidery. These ornamental gardens were actually best viewed from the house, where guests could look down upon the complex patterns and be impressed by the human control and the presentation of the order of nature.[8]

Thus, French gardens did not reflect their surrounding environment, but rather presented an abstract structuring of nature that became an art form. The garden was an area where human intelligence could reveal nature's hidden order.

Until the 1700s, English gardens closely imitated these formal gardens. In the eighteenth century, previous views of nature began to shift, with the English rivaling the French in their role as landscapers. Like its French counterpart, the English landscape reflected political and cultural tenets that governed eighteenth-century English society. Gardens designed in England began to vary from the typical seventeenth-century French formal garden, developing a new aesthetic sprung from a major philosophical revision of the idea of nature.[9]

The invention of the sunken fence about 1690 allowed landscape gardeners to create the perception that the estate extended to the horizon unimpeded by fences. The French may have invented these structures, but the English used them to shape a new English landscape. These sunken fences ran like trenches, perpendicular to the line of view and, when placed correctly, created an invisible barrier between the estate boundary and the cultivated or wild landscape on the other side. The exclamation of surprise people gave when they came upon these barriers unexpectedly, supposedly earned them their name: *ha-ha* (figure 6). The deep, wide ha-ha kept grazing animals from wandering off the property and brought unimpeded distant views into the garden, giving the impression that the family estate extended to the horizon.[10]

English landscape gardeners carried this idea to an extreme that was limited only by money and by site-specific constraints and opportunities. Horace Walpole, the eighteenth-century English landscape historian, lauded William Kent (1685–1748),[11] the preeminent English landscape gardener who made extensive use of the ha-ha fence. According to Walpole, Kent "leaped the fence, and saw that all nature was a garden."[12] Kent's designs created an image of space that exceeded the designs of his predecessors. Kent was inspired in part by nature as depicted in eighteenth-century landscape paintings. Painters, poets, and landscapers frequently made reference to a classical Arcadia, part of ancient Greece, invoked as a mythical reminder of a simple, pastoral life where humans and nature lived in harmony. In the late eighteenth century, an industrializing society again turned to the myth of Arcadia as an antidote to the gradual urbanization and mechanization of life.[13]

Figure 6. A sunken fence or ha-ha. Reproduced with permission from Anthony Huxley, *An Illustrated History of Gardening* (London: Paddington Press, 1978), 97. Illustration by John James, 1712 (?).

This flourishing landscape art movement rode the coattails of a transformation in what was considered nature but what was in reality just another human-made landscape. Landscapers skillfully engineered compositions of nature, creating vistas out of rocks, trees, and water, erasing all marks of human activity to produce peaceful pictures of nature. As cities grew increasingly polluted and disease-ridden, landscapers transformed the countryside for the elite, often razing entire villages[14] in order to create a view of nature—a nature presumably devoid of human intervention. Taking center stage in this theater was the grass plant (see box). Using grass, landscapers blended estates into the property surrounding them, producing an illusion that their created landscapes flowed smoothly into nature.

Grass plants (figure 7) grow from a crown near the soil's surface and produce stems (composed of leaf bases) and blades (leaves). A lawn mower or grazing animal cuts off the blades, but crowns remain intact and are able to produce more blades. If mowing or grazing ceases, grass plants produce flowering stems that produce seeds. Grass invades surrounding areas by producing new plants from seeds or from the growing points, or by sending out rhizomes. Lawn grass is but one of the ways wild grasses have been domesticated; much of the world's food comes from domesticated grasses such as wheat, rice, barley, oats, rye, sugarcane, millet, bamboo, sorghum, corn, and pasture grasses. Although various grasses thrive in a range of environmental conditions, all share unique biological characteristics.

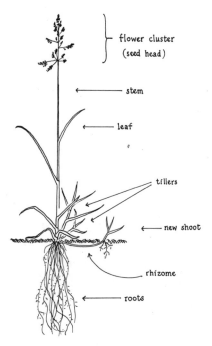

Figure 7. The grass plant. © Drawing by Lauren Brown.

Though Kent and other landscape artists contributed to the establishment of the lawn as the essential component of the English landscape, it was Lancelot "Capability" Brown who brought the lawn to its full prominence.[15]

Brown, a landscape artist who earned his nickname from his habit of discussing a particular site's "capabilities," would frequently obliterate the work of his predecessors, hiring gardeners to contour land into perfectly smooth surfaces on which sweeping lawns would be planted, thus giving these landowners their own expansive vista of "Nature." Brown's landscapes thrived as gardeners scythed, brushed, and swept their lush green surfaces, rolling the verdant carpets to smooth out any irregularities (plate 3).

Brown rarely created completely flat areas, but rather shaped the ground into a concave or convex surface to focus an observer's view in a particular direction. Unlike Kent, Brown avoided creating painterly scenes that invoked images of Arcadia, preferring simpler open spaces and smooth lines.

Indeed, Brown's penchant for planes of grass seems to have had no bounds. In the name of Nature, Brown destroyed yew hedges, displaced villages, dismantled houses, and chopped old avenues of mature trees to the ground. In Moor Park, he specified that an old garden be uprooted in order to plant numerous lawns that soon became famous all over England. Scottish landscaper Ian Hamilton Finlay has said of him: "Brown made water appear as Water, and lawn as Lawn."[16]

The success of Brown's landscapes became the keystone of a revolution in aesthetics that cemented the lawn as the great icon of late eighteenth-century British society. It cannot be overemphasized how closely Brown's success was tied to the English climate. England is a country of mild winters, moderate temperatures, and high humidity—all conditions favorable to the growth of the grasses he planted so abundantly. The lawn soon became a symbol whose roots reached across the Atlantic to different, less hospitable, environments.

History of the Lawn in the New World

When the British colonized the New World, they brought with them their ideas about nature. Long-ago cultivated, grazed, or developed, the island of Great Britain had no unaltered landscapes left. British landscape design that "improved" a natural landscape in reality only added new artificial dimensions to an existing landscape already highly modified by humans. In the New World, colonists were not prepared to cope with seemingly infinite expanses of wild land inhabited by unfamiliar plants, animals, and people.

The British also brought a strong preference for an English landscape to the New World. Although new climates and totally different vegetation and soils challenged English aesthetics and life-styles, the immigrants clung to practices and ideals of their cultural past. Vast expanses of wild land ready for farming only fueled this image.

Grass played a central role in English agriculture since it sustained sheep and cattle. William Wood, an English traveler, warned settlers coming to New England in the 1630s to seek places where there would be enough grass to feed cattle. In general, grazing land was scarce; domesticated animals quickly ran out of grass, and forestland then had to be cleared to create new pastures. Seed brought from Europe was used, and soon European plants such as bluegrass and white clover, which were adapted to the harsh requirements of pastoralism, began to take over wherever cattle grazed. By 1640 a regular market in European seed existed in Rhode Island, and within one or two generations these plants had become so common they were regarded as native. During the eighteenth century, European pasture plants including timothy and fowl-meadow grass and the legumes, red clover and alfalfa, were common on pastures throughout the colonies.[17] Thus the pasture, a mixture of grasses, legumes, and other plants, became a common landscape feature in much of colonial America. Pasture was not only a rural feature but was also seen in many villages and towns, where it clothed the village common.

One of the many travelers who brought the English landscape back to the United States was Thomas Jefferson. Jefferson visited and admired some of the most important English gardens, and he was clearly struck by the pastoral image of a classical building set in a field of green. "Moor Park . . . the lawn about thirty acres" is recorded in his notebook.[18] He later incorporated this vision in the design of Monticello, where he used the classical precedents of villas, English landscape ideas centered around the lawn, and American ideals of independent citizen-farmers. The view from his estate began at the mansion, flowed through a lawned garden and woods to the view of productive farm landscape, and ended with the wild, unsubdued scenery on the fringes of the estate.

About this time he created a place for the education of his citizen-farmers: the University of Virginia. Here the buildings were organized in the shape of a U surrounding a tiered lawn. In Jefferson's original design, one was to look out on nature in the form of a botanical garden, which was never built, and beyond to the distant hills. The whole Jeffersonian complex of the University of Virginia is referred to as "The Lawn," basing its name on the beautiful lawned space contained within it (figure 8).

Lawns were not uncommon in America at the turn of the eighteenth century, but with few exceptions they were kept to a minimum, the grass cut by hand scythes or kept short by grazing animals. At Mount Vernon, George Washington created a great tree-lined swath of grass sweeping down to the Potomac beyond the ha-ha that separated it from the smaller lawn around the

Figure 8. The Jeffersonian complex at the University of Virginia centers around a tiered lawn and is known to thousands of graduates as "The Lawn." Photograph by Diana Balmori.

house. Cutting this expanse of grass was a cause of concern until he hit on the happy idea of using browsing deer to keep it short.[19] Much more common than lawns were treeless, shrubless, rather unkempt weedy properties whose front yards, especially in the South, were tidy patches of swept bare ground with occasional planting beds or shrubs.

The Suburban Lawn

When did the mowed lawn take such firm root among the majority of American households? According to Kenneth Jackson's classic, *The Crabgrass Frontier: The Suburbanization of the United States,* the mowed lawn appeared in the mid-nineteenth century along with the creation of the American suburb.[20] The Industrial Revolution had transformed American cities, and a new middle class began to seek residences outside cities. Parks, cemeteries, and suburban cottages were recommended for their aesthetic, moral, and recreational benefits as well as their contribution to physical health. Builders landscaped green areas around houses to satisfy their clients who believed green plants created reservoirs of clean air and healthy home environments. By 1870, detached housing had emerged as the suburban style of choice, with drawings typically depicting an isolated structure surrounded by a yard.

The lawn, carrying the English connotations of nature with it, became a symbol of prestige in the nineteenth-century suburbs. Similarly, centers of towns in New England, the old "commons," which had provided the setting for many useful activities such as rope making, hay growing, military drills, and town fairs in the eighteenth century were transformed from bare stamped earth, cultivated fields, or cemetery grounds into lawned and treed parks, now called "greens," in the nineteenth century. With this grassy transformation, most of the economic functions of the commons were shifted to other places. At least one American

Plate 1. The medieval garden's lawn was full of small flowers and a mixture of grasses kept short by use. The medieval lawn, an early ancestor of the modern lawn, was known as the flowery meade. Anonymous German master, 1410, *Das Paradiesgärtlein,* Städelsches Kunstinstitut Frankfurt am Main. Photograph © Blauel Kunst-Dia.

Plate 2. Productive and ornamental garden, Château de Villandry, France. This French garden has panels of grass as a background to the planted edges and to the perfect figures of fountains and parterres. The crushed stone paths, rather than the panels of grass, are for walking. Photo: DiaFrance.

Plate 3. Moor Park, one of the most cited of Capability Brown's landscape designs, is described by Thomas Jefferson as "thirty acres of lawn." Richard Wilson, *Moor Park*. © Marquess of Zetland. Reproduced in Roger Turner, *Capability Brown and the Eighteenth Century English Landscape* (New York: Rizzoli, 1985), 85.

Plate 4. A New England green. Photograph by Diana Balmori.

writer, J. B. Jackson, has condemned this transformation of true "commons" into "greens" (plate 4).[21]

The upkeep of green spaces required intensive labor, and early suburbanites depended on hired labor or on sheep or deer to keep things tidy. The advent of the lawn as the common people's art form would be made possible by technology: in 1830, the Englishman Edwin Budding invented the lawn mower (figure 9). As these machines became available over the following decades, modest householders could own these mowers and keep their lawns tidily cut without the help of gardeners or flocks of sheep. When applying to the British Patent Office, Budding claimed: "Country gentlemen will find in using my machine an amusing, useful and healthful exercise."[22] Jane Loudon heralded the invention of the lawn mower as "a substitute for mowing with the scythe . . . particularly adapted for amateurs . . . but it is proper to observe that many gardeners are prejudiced against it."[23] Forty years after the lawn mower's introduction, Beeton's *Dictionary of Gardening* describes the use of mowers as "too well known to need description. . . . These useful machines are fast supplanting the scythe both on large and small lawns."[24]

Finally, the well-manicured lawn was within easy reach of average citizens. Whether they had an acre or a tiny patch of land, they could tend it themselves and create whatever image they chose for their surroundings.

In the 1840s, two Americans, Alexander Jackson Davis and Andrew Jackson Downing, vastly extended the popularity of the small detached houses of England in the United States, publishing highly successful books containing numerous illustrations of bungalows surrounded by lawns and gardens. Downing was convinced that the environment directly affected behavior. His 1841 book, *A Treatise on the Theory and Practice of Landscape Gardening, Adapted to North America,* discussed the principles of landscape design founded in natural scenery.[25] This book, widely read, encouraged Downing to write several more, most

Figure 9. Budding's invention of the lawn mower quickly replaced the scythes and brooms once used to maintain the lawn. William Wollett, *West Wycombe*, 1770 (detail). Copyright British Museum. Early advertisement for the lawn mower reproduced with permission from University of Reading, Institute of Agricultural History and Museum of English Rural Life.

with the theme of encouraging Americans to improve their surroundings, though his ideas were all based on English landscaping principles. A contemporary commented that "the value of
Downing's books has been great, not because of their technical
excellence, for they are very poor in quality, but because they are
full of life and interest. It is the man not the architect that wins
the popular ear, and compels his readers to allow that the subject
is entertaining and enjoyable."[26] From its publication in 1841
until the end of the century, Downing's *Treatise* remained the
average homeowner's standard reference. In it, the lawn was the
unifying theme (figure 10).

Separateness, Downing pointed out, had become essential to
the character of the suburb. As Kenneth Jackson writes, "Although visually open to the street, the lawn was a barrier, a kind
of verdant moat separating the household from the threats and
temptations of the city. It served as a means of transition from
the public street to the very private house. . . . The sweeping
lawn helped civilize the wild vista beyond and provided a carpet
for new outdoor activities such as croquet, a lawn game imported from England in the 1860s, tennis and social gatherings"
(figure 11).[27]

In New York and other large cities, zoning laws made their
appearance and twenty-five-foot setbacks from the street became standard;[28] these distances seemed monumental compared to old row houses that abutted directly on the street. On
the city outskirts, legal covenants requiring structures to be set
back from the street by a minimum number of feet were written
into many property deeds from the 1880s on. "The idealization of
the home as a kind of Edenic retreat . . . where the family could
focus inward . . . led to an emphasis on the garden and lawn,"
wrote Jackson.[29]

The lawn became the symbol of suburbia, championed by
Frederick Law Olmsted, the famous landscaper of New York
City's Central Park and many suburbs, who viewed the lawn

Figure 10. Andrew Jackson Downing's landscapes, as shown here, incorporated trees, lawns, and gravel walks. From Andrew Jackson Downing, *A Treatise on the Theory and Practice of Landscape Gardening*, 1. Originally published 1841. Reprinted 1977 (9th ed. of 1875) by

Figure 11. The Tennis Party. By the second half of the nineteenth century, lawn games became a feature of middle-class suburban homes. Sir John Lavery (1856–1941) *The Tennis Party*, 1885 (detail). City of Aberdeen Art Gallery and Museums Collections.

as a sort of community parkland. For Olmsted, the front lawn of a house in a suburb unified the whole residential composition into one neighborhood, giving a sense of ampleness, greenness, and community.[30]

The curvilinear layout of American residential streets, with houses set well back from the road behind front lawns with informal plantings of trees and shrubs, a uniquely American residential characteristic, was first proposed and built for an industrialist by Andrew Jackson Davis in his Llewellyn suburb twelve miles west of Manhattan.[31] It was popularized by Downing, Olmsted, and others. Olmsted and Vaux's plan for Riverside near Chicago has become an archetype of this American suburb (figure 12).[32]

The lawn continues to have strong philosophical advocates today, although there is a shift in locale. William Whyte, sociologist and urban advocate, sees the lawn as important to modern American cities. In his book *City: Rediscovering the Center* he writes: "A salute to grass is in order. It is a wonderfully adaptable substance, and while it is not the most comfortable seating, it is fine for napping, sunbathing, picnicking, and Frisbee throwing. Like movable chairs, it also has the great advantage of offering people the widest possible choice of seating arrangements. . . . Grass offers a psychological benefit as well. A patch of green gives us a refreshing counter to granite and concrete, and when people are asked what they would like to see in a park, trees and grass are usually at the top of the list."[33]

The Lawn's Conquest of North America

The lawn, since its invasion of the Eastern Seaboard nearly three centuries ago, has spread to every corner of North America. This is a bit hard to comprehend, since the lawn arose in the mild, moist, climate of England, whereas North America is a continent with many harsh climates and an enormous diversity in vegetation and soils. Climates range from the extreme winter cold of

Figure 12. Plan for Riverside, Illinois, by Olmsted and Vaux, 1869. Winding streets with continuous front yards of lawns and trees defined the character of the new American suburbs. Curving streets and green lawns declared that the homes were "in the country." Olmsted, Vaux and Co., 1869 Oak Forest/Riverside Plan, *Journal of the Society of Architectural Historians* 42, no. 2 (May 1983): 156.

Canada and the northern United States to the intensely hot and droughty summers of the South, from the extremely wet Northwest Coast, to the wet East Coast, to the extremely dry Southwest. These climatic patterns were reflected in the great range of natural vegetation the first explorers encountered when they painstakingly boated and walked over the surface of this vast continent. Evergreen forests crossed the continent in the North, deciduous forests covered the great stretch of land from the Mississippi eastward, native grasslands, not English pasture grasses, swept northward from the Gulf of Mexico, west of the Mississippi and east of the Rockies into Canada, and in the Southwest large areas of desert vegetation occupied lowlands that ran well into Mexico (figure 13). Most of these vegetation types and climates were not in the least hospitable to the English lawn. Nevertheless, the same continuous sward of green front lawn joins block after block, whether on Locust Street in Champaign-Urbana, Illinois, or Ocean Front Boulevard in Santa Monica, California. Just how could the mild-mannered lawn conquer these hostile environments?

The adaptation of the lawn as the green mantle which unified the suburbs was initially made possible by Budding's invention of the lawn mower. Budding's invention was followed by many others and paralleled by an explosion in agricultural knowledge and technology. We learned about plant nutrition, about fertilizers and how to produce them, about plant disease and insect depredations of plants and how to combat them, about water requirements of plants and how to meet them with irrigation, and, perhaps most important, how to breed plants better adapted to particular environments.

Technology soon found its way into the development of the lawn care industry and through entrepreneurialism became relatively inexpensive and available to homeowners through retail outlets and professional lawn care companies. Grass seed companies developed varieties especially for the noncommercial

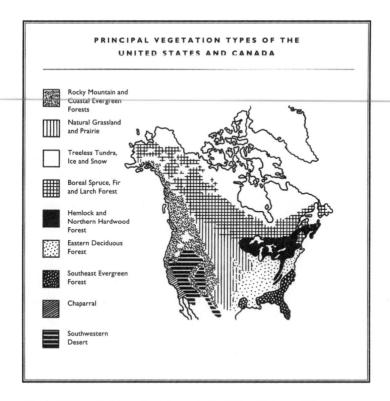

PRINCIPAL VEGETATION TYPES OF THE
UNITED STATES AND CANADA

Rocky Mountain and
Coastal Evergreen
Forests

Natural Grassland
and Prairie

Treeless Tundra,
Ice and Snow

Boreal Spruce, Fir
and Larch Forest

Hemlock and
Northern Hardwood
Forest

Eastern Deciduous
Forest

Southeast Evergreen
Forest

Chaparral

Southwestern
Desert

Figure 13. Principal type of vegetation in the United States and Canada.
Adapted from Henry J. Oosting, *A Study of Plant Communities: An
Introduction to Plant Ecology,* 2d ed. (San Francisco: W. H. Freeman and Co.,
1956), 271.

market. Chemical companies responded to the needs of small-
scale purchasers. Sod companies made the lawn shippable and
made possible instant lawns. And thus the lawn was inexorably
commercialized and through entrepreneurialism was able to
overcome naturally occurring environmental barriers. Not only
because it was possible but also because it was preferred and
liked by a mass audience conditioned by a long cultural history,
the lawn became the landscape of choice from Kennebunkport,
Maine, to Las Vegas, Nevada.

But there was a price to pay.

Chapter 2
Questioning the Lawn

Millions of Americans love their lawns and are satisfied with a rich sward of pure grass; but others have lawns that depart from this standard. These Americans have begun to question the purpose of a lawn and its social, environmental, and economic costs. Many are starting to break with suburban orthodoxy and to reconsider the management and design of the space around their homes. In essence they challenge the monolithic view of the American lawn.

We begin by describing the nonconventional lawns of four homeowners in Maryland, Connecticut, and Georgia. Examining their dissent helps us understand what the American lawn is and what it may become. In the remainder of this chapter, we explore the rise in environmental awareness that has led to growing opposition to the traditional lawn.

Walter and Nancy Stewart of Potomac, Maryland

Walter Stewart did not look forward to cutting the lawn, but when the weekend came around, he knew it had to be done.[1] But one day was different: the lawn mower failed to start. What better excuse than a broken lawn mower to placate one's conscience! As time went by, the mower remained broken and the grass grew longer and longer; gradually it developed into a meadow. After a time, the Stewarts came to the conclusion that nothing except convention dictated that they must have a front yard of mowed grass. "People may think we are lazy, but who cares." This is precisely what their neighbors first thought, but soon their tolerance for this eccentric behavior turned to impatience. The Stewarts were informed that the neighborhood was not pleased with their landscape design; the neighbors insisted that a conventional lawn was more appropriate. The Stewarts responded with a lengthy letter explaining why they chose to let their lawn grow into a meadow. They proposed that a meadow provided more than a lawn could ever hope to provide: a haven for animals; food for countless organisms; privacy for the homeowner; and, simply, a more "natural" environment. They also noted that maintaining a conventional lawn was a waste of their time and energy, since it takes considerable energy to keep a lawn from turning into a meadow in Maryland.

Walter and Nancy Stewart received a citation from Montgomery County stating that their lawn was in violation of the twelve-inch rule and requiring the lawn to be cut within ten days. Any lawn that exceeded twelve inches in height was considered a municipal health risk. The Stewarts defended their choice of a meadow on ecological grounds and challenged the validity of the county's health risk ordinance. In the end, the county changed its regulations, thereby allowing the Stewarts to keep their meadow.

Michael Pollan of Cornwall, Connecticut

Michael Pollan is a homeowner who espouses a new sense of nature. He has chosen to deviate further and further from the green carpets of uniform height that conventionally connect homes in the suburbs. When Pollan first bought his home, he faithfully trimmed his lawn. Although not his favorite pastime, he did gain a measure of enjoyment from being outdoors and caring for his grass. From endless weeks of mowing, he came to know his yard intimately. He knew when bumps were coming, where the mower would feel the strain of an especially thick patch of grass, and when he was only five minutes away from completion. He recognized that crabgrass grew in drier areas and clover preferred depressions where water tended to collect.

With time the regimen of mowing lost its appeal and other aspects of his yard became far more fascinating. Pollan developed a particular liking for gardening, for growing vegetables and fruits. With time, fruits and vegetables became a significant part of his landscape design. Fruit trees were planted in his front yard, and by trial and error he learned the ecological requirements of various kinds of vegetables. There was a subtle pleasure in finding the right balance of water, nutrients, and light that would enhance growth. All of these pursuits he found to be greatly preferable to mowing. One does not learn much from the grass that one mows. Even the word *mow* conjures up images of conquest and submission to indiscriminate assault. In *Second Nature,* an account of his experiences in the garden, he remarks: "Gardening, as compared to lawn care, tutors us in nature's ways, fostering an ethic of give-and-take with respect to the land. Gardens instruct us in the particularities of place. They lessen our dependence on distant sources of energy, technology, [and] food . . ."[2]

And so, Michael Pollan continues to extend his gardens. His turf gets smaller and smaller; he has bordered his yard with a hedge, breaking the continuous carpet from home to home. And he has a half-acre meadow of black-eyed Susans and oxeye daisies that yield beautiful flowers from early summer to the time of killing frost.

Murray and Ann Blum of Athens, Georgia

When Murray and Ann Blum built their house in Athens, Georgia, they insisted that the developer leave all of the trees already growing on the property.[3] Between these relatively mature trees, the Blums added sycamores, a Chinese walnut, and French locusts to provide beauty throughout the year. Murray Blum's lifelong interest in insects led them to also add flowering shrubs to provide dependable daily resources for insects for much of the growing season.

When the woody plants matured, the yard looked more and more "natural." Plants and animals found their way into the mix that the Blums provided. As the interactions between fauna and flora multiplied, the Blums realized that they were "giving something back to nature." Their front yard had become a part of nature again (figure 14).

"So we let things grow. We've let the ground covers work out their territories. The recent invasion of *Vinca* has provided the elements of a major battle as it approaches the English ivy." Nature has not been left totally to its own devises, however. Murray has great fun finding plants, especially wildflowers, to add.

The neighbors have not been happy with all of these activities. The Blum's yard does not fit in; it interrupts the green continuity of the street's front lawns. To appease the neighbors and to legitimize this one-acre irregularity, the town declared a one-acre bird sanctuary and installed a sign on a large tree.

"Grass cultivation is energy-intensive, consuming vast amounts of resources, e.g., fertilizer, biocides, and water. Grass seems to us an expensive anachronism in a world where people are starving. For us, the beauty of wildflowers and their pollinators far surpass the monotony of lawns. Our yard is no place to play baseball, but that is a small price to pay for the medley of trees, shrubs, thickets, leaves and flowers and for the omnipresent birds, bees, and bugs."

Starting with a desire to protect the trees growing on their newly purchased lot, the Blums have given their yard back to nature and at the same time have reduced their dependence on scarce resources. In their own front yard they are able to express

THE YARD
FROM HELL

Figure 14. Murray Blum in his front yard in Athens, Georgia. On May 6, 1992, the *Atlanta Journal* featured a story about the Blums and their yard. © Photograph by William Berry, 1992. Reprinted with permission from the *Atlanta Journal* and *Atlanta Constitution*.

their aesthetic preferences and their concern to promote harmony between humans and nature.

Joel Meisel of Prospect, Connecticut

Joel Meisel's dissent developed gradually, much as Michael Pollan's did.[4] Meisel noticed that wildflowers were continually invading his lawn. These were not simply dandelions or plantains but meadow flowers. As a scientist concerned with plants, Meisel's appreciation went beyond simple delight in beauty. *Krigia virginica,* a small, yellow-flowered plant had its own natural history which he understood. He enjoyed the fact that these plants had found their way to his front yard and found himself spending idle moments contemplating the marvel of their growth, survival, and spread. He continued to mow his lawn, but he avoided the patches of wildflowers; soon his lawn was a patchwork of lawn and flowers.

His neighbors were concerned about his increasing deviation from the perfect lawn. Inquiries were made to determine if something was wrong. To the more curious, Meisel explained his interest in the little reproductive miracles that were taking place in his lawn. He found that some neighbors were impressed with his knowledge of what was going on in his lawn and, consequently, respectful of his dissent. Soon other people in the neighborhood were growing wildflowers, although not in the "lawn." Meisel continues to mow around hawkweed, mullein, and oxeye daisies, and he and a next-door neighbor share a wildflower garden between their homes.

Why Do Some People Wish to Change the Lawn?

The examples cited above could be multiplied manyfold, from Arizonans who prefer a patch of desert to a carpet of green to North Carolinians who value a longleaf pine forest more than a lawn. What message is there in this dissent? To answer this question we need to explore four elements of varying importance: time, changing aesthetics, our desire to experiment, and environmental consciousness.

Clearly, many people dislike the burden that constant lawn care places upon them. These days, time is often in short supply. In the minds of some, lawn care can be a demanding taskmaster. The inexorable routine of weekly mowing, especially under the impetus of added fertilizers and irrigation, can become a boring and time-consuming routine. Some lawn owners find comfort in developing lawn management strategies that demand less time and energy. Others may hire a lawn care company that, for a price, relieves the owner of most responsibility.

In a very direct way, the experiences of the Stuarts, Michael Pollan, the Blums, and Joel Meisel reflect the beginnings of a change in landscape aesthetics. The eighteenth-century vision of a distant nature has lost its strength. In its stead, a sense of kinship and curiosity about living things is growing, and there are attempts to represent this more inclusive vision in the lawn and garden, for this is still where most daily contact with other living things takes place. In time, these efforts taken in conjunction with artistic visions will shape this new sensitivity into a different aesthetic image of the landscape.

As a people, Americans are explorative! Our history is filled with tinkering, innovation, and invention. The lawn offers an arena for experimentation. One of the most alluring aspects of "lawn tinkering" involves, not tinkering with technology, but tinkering with nature. The Stewarts became fascinated with a natural process called plant succession by which their lawn became a meadow by naturally occurring processes. Michael Pollan became fascinated with the potential of his land to produce things other than lawn grass, and Joel Meisel decided to use part of his lawn to study the behavior of wildflowers. The Blums saw in their yard an opportunity to give something back to nature. It seems clear that the interplay between imagination and exploration play a large part in the dissent from the green carpet lawn. This impetus for innovation is aided by the examples we see around us, so the more people who innovate, the more examples we will all have.

The Rise of Environmental Consciousness

A final reason for dissent from conventional concepts of the lawn is the rise of environmental consciousness. The following overview touches only the surface of a deeply philosophical and complex subject.

For much of human history, there was great fear of the unknown. The unknown was everything out there. Indeed, nature had much that could thwart and humble humans: wild beasts, floods, earthquakes, plagues, wildfires, crop disasters, sickness, unexpected frosts, droughts, locusts, erosion, hurricanes, and dozens of other natural events. Succor was centered on relationships between people, since these formed a bulwark against the unpredictability of nature. Not unexpectedly, this perception of the relationship between people and their animate and inanimate surroundings led to a philosophy of "man apart from nature." Under this paradigm, humans viewed nature with some degree of fear but also as something to be manipulated in what were perceived to be the best interests of humanity. The philosophy of the eighteenth-century landscape designers reflected this view. For the upper classes, the pleasure garden and the lawn were considered human buffers against wild nature.

At about the time the lawn came into fashion, Europeans had gone far toward subduing nature. The demands of an increased population had changed Europe into a largely human-controlled landscape. Wilderness as we think of it was essentially gone. European culture was also beginning to feel the effects of the Industrial Revolution, however. In this world of machines, of crowded and disease-ridden industrial cities full of chimneys belching black smoke, the beautiful image of serene, grass-carpeted landscapes in rural areas unmarred by urbanization became a social ideal. The rural landscape, even though it was itself a product of human manipulation, was seen as nature when juxtaposed to the grim view of the human-created urban environment. The natural world of trees, mountains, pastures, grazing cattle, streams and rivers, and rocks was contrasted to the urbanized and industrialized works of humankind.

As northern Europeans continued to emigrate to North America, they carried with them this eighteenth-century philosophy. These new emigrants viewed the great expanses of wilderness through the eyes of Europeans whose landscapes had long been under human control; they looked through the prism of "man apart from nature." Wilderness was considered an impediment, a source of danger from savages, wild animals, and the unknown. Wilderness needed to be cleared as quickly as possible to minimize these inherent dangers and to convert it to purposes useful to people. Nature was to be subdued and made the servant of humankind.

Although "man apart from nature" has a history that stretches through the millennia, the idea has reached its zenith in the twentieth century. The extraordinary accomplishments of the industrial age with its fantastic advances in science, medicine, space exploration, engineering, computing, agricultural and industrial production, transportation, and communication have led many to believe that the work of men and women can solve any problem. As the old Navy saying goes, "The improbable we do immediately, the impossible takes a little longer." Powerful individuals in the fields of science, technology, finance, economics, industry, and politics firmly believe in the "technological fix": that human ingenuity, technology, and organization can do anything, that humankind knows no bounds. This kind of thinking, so common in today's world, represents the ultimate in "man apart from nature."

Doubts about "Man Apart from Nature"

The idea that people are above the laws of nature has been questioned for a long time. More than four hundred years ago, Sir Francis Bacon observed that "Nature, to be commanded, must be obeyed." This aphorism has direct bearing on how we design and manage our lawns. For if we are *not* above the laws of nature, some of what we do to the lawn may change naturally occurring processes in such a way that nature's response will be inimicable to humankind's long-term best interests. In other words, we cannot indiscriminately manipulate our lawns without in some measure diminishing our local, regional, and global environment. This is perhaps the major reason many people are troubled by the conventional view of the lawn as a lush green carpet of Kentucky bluegrass.

Growth of Scientific Knowledge

A great body of scientific knowledge has developed over the centuries that lends strong support to the idea that humans are not unique, at least in terms of the natural processes that govern the earth and the stars.

Today science accepts the existence of a chemical basis for life. This chemistry underlies how we become ill, how we get better, how medicines work, how plants make food, and how organisms decompose. It is believed that every process in the biological world can be described in terms of chemical reactions. Although there is still a vast amount to learn, biochemists working with a huge range of organisms have demonstrated that many chemical processes are common to all organisms including humans.

Inviolable forces are at work in the atoms that make up our bodies, and forces unimaginable to most of us are at work in the universe. The physics and chemistry of humans are not unique; our bodies are composed of matter that obeys describable laws in a universe so vast that humanity is nothing but a speck.

Other notions of human uniqueness were shaken with the publication of Charles Darwin's *Origin of Species*.[5] Darwin's ideas dramatically changed the way people saw themselves in the natural world. Today many people have come to accept that humanity shares a common lineage with other primates and with all other forms of life. We are not separate and distinct. Whales, elephants, mice, and humans are just different products of time and evolutionary processes.

Ecologists and other scientists who study whole forests, lakes, oceans, and the atmosphere have demonstrated a wide array of processes that are essential to maintaining life on the earth. These are processes such as those that maintain the composition of the atmosphere, that underlie the constant cycling of nutrients between living and dead matter, that maintain the soil, and processes that govern the cycling of water. All life depends on these processes.

The scientific evidence seems overwhelming that humans are part of the fabric of nature and cannot be separated from it. We are living, breathing organisms subject to gravity, chemical bonds, energy flows, and nutrient cycles. We have a common lineage with other life-forms with whom we share common biological processes, and like other organisms we depend on naturally occurring processes to maintain conditions favorable to life. The new motto should be that the only things in this world you have to do are get born, die, pay taxes, and *abide by nature*.

Development of Environmental Thought

Environmental thinking challenges the idea that humans are above nature and that nature can be thought of simply as a commodity to be bought, sold, altered, or destroyed. Environmental thinking sees all life as part of nature and believes that our manipulations of the earth's surface should be carried out with the deepest respect for nature lest nature respond with an irrevocable altering of the very earthly conditions upon which humankind is dependent for survival.

The roots of environmental thought are deep and books about it fill many shelves. To catch the historical flavor of its evolution, it is useful to touch on the contributions of a few of its thinkers. In 1851 Henry David Thoreau wrote: "At present, in this vicinity, the best part of the land is not private property; the landscape is not owned, and the walker enjoys comparative freedom. But possibly the day will come when it will be partitioned off into so-called pleasure-grounds, in which a few will take a narrow and exclusive pleasure only,—when fences shall be multiplied, and man-traps and other engines invented to confine men to the *public* road, and walking over the surface of God's earth shall be construed to mean trespassing on some gentleman's grounds. . . . Hope and the future for me are not in lawns and cultivated fields, not in towns and cities, but in the impervious and quaking swamps. . . . Give me the ocean, the desert or the wilderness!"[6] Thoreau's remarks were as spiritually oriented as they were foresighted. Although written a century and a half ago, he expresses respect for nature and captures America's twentieth-century concern to protect nature and pristine environments.

In his timeless book *Man and Nature: Physical Geography as Modified by Human Action,* George Perkins Marsh in 1864 described the danger humanity faced if it continued to raze the forests of the world. The issues set forth there are still pertinent today and will be into the foreseeable future. Marsh chronicled the strong correlation between the destruction of woodland resources and the collapse of human empires and wrote in scathing terms of land management practices of his day. "Man is everywhere a disturbing agent. Wherever he plants his foot, the harmonies of nature are turned to discords. The proportions and accommodations which insured the stability of existing arrangements are overthrown. Indigenous vegetable and animal species

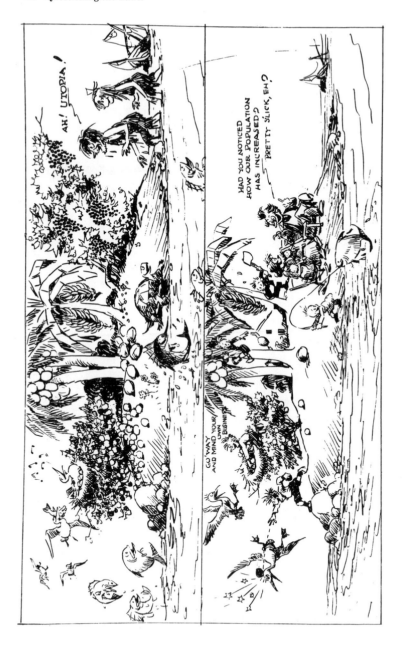

Figure 15. Ding Darling, famous political cartoonist and early environmentalist, captures the widely held belief that humans could destroy some part of the earth and simply move on to another in his 1936 cartoon of utopia. The cartoon is dedicated to Dr. Paul Sears, eminent ecologist of the mid–twentieth century and author of the renowned book on the great North American drought of the 1930s, *Deserts on the March*. Cartoon property of F. H. Bormann.

are extirpated, and supplanted by others of foreign origin, spontaneous production is forbidden or restricted, and the face of the earth is either laid bare or covered with a new and reluctant growth of vegetable forms, and with alien tribes of animal life. These intentional changes and substitutions constitute, indeed, great revolutions; but vast as is their magnitude and importance, they are . . . insignificant in comparison with the contingent and unsought results which have flowed from them."[7] Thus, Marsh clearly and forcefully saw the unanticipated side effects of nineteenth-century technology, a problem that has grown to enormous proportions in today's world.

In 1948, William Vogt wrote *Road to Survival*.[8] Among his provocative ideas were the notions that technology makes the world effectively smaller and that localized acts can affect the entire planet. Vogt challenged the concept of limitless space which held that humans could destroy one area and move to another (figure 15), or, in a modern context, that we could discharge our hazardous wastes knowing that almost infinite dilution would render them harmless, or that we could store them in out of the way places where they were sealed off from human society. Vogt challenged the widely accepted notion that more growth was always better and that technological advancement was an inherent and unequivocal good.

In 1949, Aldo Leopold's book *A Sand County Almanac*[9] described humanity's role in light of all life around us. The book articulated the connection between preserving wilderness and preserving all nature including ourselves. Arguing from a strong foundation in both science and ethics, Leopold contended that we were dooming ourselves by disrupting nature. Most humans imagine that they are sustained by economy and industry. Leopold did not deny this fact, but he pointed out that what sustains economy and industry is all living things.[10] In essence, he told us that too narrow a focus in our economic pursuits can result in unanticipated responses by nature that could prove dangerous to human society.

In her stunning book *Silent Spring,* Rachel Carson (figure 16) made the American public keenly aware that our personal health was being compromised by our general neglect of ecological connections.[11] In her indictment of the use and overuse of chemical pesticides in America, she argued that our culture had headed

Figure 16. Rachel Carson was among the first scientists to alert the American public about the unappreciated and dangerous side effects of the chemical age. © Photograph by Erich Hartmann, used by permission of Rachel Carson Council, Inc.

willy-nilly into the age of chemicals without considering the to-
tality of nature; processes within nature were working to undo
the very things we hoped to achieve using pesticides, thereby
causing large-scale imbalances. In our manipulations of nature,
she urged us to seek to use the processes and inviolable realities
of nature to our advantage and to avoid the use of pesticides.

The initial public response to *Silent Spring* was an instinc-
tively corporeal one, concern for human health, rather than an
understanding of the broader implications of her work: the need
to know our place in ecology. Yet this response was natural when
one reads some of the passages from *Silent Spring:* "In one of the
most tragic cases of endrin poisoning there was no apparent
carelessness; efforts had been made to take precautions appar-
ently considered adequate. A year-old child had been taken by
his American parents to live in Venezuela. There were cock-
roaches in the house to which they moved, and after a few days a
spray containing endrin was used. The baby and the small family
dog were taken out of the house before the spraying was done
about nine o'clock one morning. After the spraying the floors
were washed. The baby and dog were returned to the house in
midafternoon. An hour or so later the dog vomited, went into
convulsions, and died. At 10 P.M. on the evening of the same day
the baby also vomited, went into convulsions . . . and [ultimately]
became little more than a vegetable."[12]

Here is a story to which all American parents can relate. This
is not the story of occupational exposure; it is the story of the
high price paid for innocent ignorance of the effect of these chemicals. The message to the public is clear: to ignore the potential
dangers of these chemicals is to place one's own children and pets
in peril, in one's own home.

Rachel Carson also demonstrated that it was not just acute
incidents that we had to fear. The accumulation of small doses of
chemicals over long periods of time constituted another source of

danger. After World War II, pesticides were being sprayed everywhere for every conceivable purpose. Carson made Americans
question the ultimate fate of these chemicals. When the smell
was gone, were the chemicals themselves gone? Carson demonstrated that these chemical poisons were often incredibly effective because they could persist in the environment for a long
time. These chemicals were designed for use in a natural setting
that had a limited ability to decompose them. If these chemicals
persisted in the environment, it could mean that there was abundant opportunity for human exposure at unlikely times in un-
suspected places. The focus came back to human health in a very
personal way. Carson writes: "The contamination of our world is
not alone a matter of mass spraying. Indeed, for most of us this is
of less importance than the innumerable small-scale exposures
to which we are subjected day by day, year after year. Like the
constant dripping of water that in turn wears away the hardest
stone, this birth-to-death contact with dangerous chemicals may
in the end prove disastrous. Each of these recurrent exposures,
no matter how slight, contributes to the progressive buildup of
chemicals in our bodies and so to cumulative poisoning. Probably
no person is immune to contact with this spreading contamination unless he lives in the most isolated situation imaginable"
(see also figure 17).[13]

Indeed, these dangerous chemicals can be found as close to
home as one's own lawn. Carson observed that "those [lawn-
keepers] who fail to make wide use of this array of lethal sprays
and dusts are by implication remiss, for almost every newspaper's gardening page . . . take[s] their use for granted."[14]

These prophets touched on many points. Marsh emphasized
that unanticipated side effects from forest destruction acting
through ecological pathways would have strong negative effects
on human societies. Vogt stressed how technology can shrink the
globe, making space less of a solution to our environmental problems. Although we are sustained by our economy and industry,

Leopold told us that these things were, in turn, sustained by nature. To injure nature was to injure ourselves. Finally Rachel Carson focused on the biogeochemical pathways of nature and how chemical contamination of these pathways can injure humans as well as other organisms and damage the ecological systems that provide the basis for society's health and well-being. Henry David Thoreau pointed out that nature has spiritual and moral values in addition to practical values. Although we have not explored this track to any degree, surely these values have a central place in any debate about nature and our use of it.

Contemporary Environmental Thought

The gateway opened by *Silent Spring* was followed by a flood of environmental literature in the late sixties and early seventies such as Barry Commoner's *The Closing Circle—Nature, Man, and Technology* (1971) and Paul Ehrlich's *The Population Bomb: Population Control or Race to Oblivion* (1968).[15] That flood has continued to this day.

One of the most important groups of environmental writers

What happens to fallout and pesticides in our environment?
Some of it decays - the rest ACCUMULATES.

It gets into plants, then plant-eaters, then meat-eaters.

All along this food chain it's CONCENTRATED because animals eat many times their own weight in food.

So the last guy in the chain gets quite a load.

WE'RE LAST GUYS.

Poisons kill other last guys, too,
making nature simpler and less stable.

Imagine a simple food chain: posies → sheep → people

In nature: SIMPLICITY = INSTABILITY

Figure 17. In 1966, George Woodwell, W. M. Malcolm, and R. H. Whittaker published the above series of cartoons in a small booklet entitled, "A-Bombs, Bugbombs, and Us." The booklet, built on the theme raised by Rachel Carson, emphasized that chemicals or radioactive materials moving through food chains affected not only plants and animals but humans at the end of the chain. Woodwell played an important role in raising environmental consciousness and was one of the founders of the important environmental organization, the Natural Resources Defense Council. From Brookhaven National Laboratories and the U.S. Department of Energy.

today can be found at the Worldwatch Institute under the leadership of Lester Brown. Some consider the institute's annual publication, *State of the World*,[16] as important and influential as *Silent Spring* and *Sand County Almanac*.

Published in twenty-three languages, *State of the World* documents the evidence, which is accumulating at an alarming rate, that unanticipated side effects of man's actions are causing the severe deterioration of natural systems. To quote Lester Brown, "Anyone who regularly reads scientific journals has to be concerned with the earth's changing physical condition. Every major indicator shows a deterioration in natural systems: forests are shrinking, deserts are expanding, croplands are losing topsoil, the stratospheric ozone layer continues to thin, greenhouse gases are accumulating, the number of plant and animal species is diminishing, air pollution has reached health-threatening levels in hundreds of cities, and damage from acid rain can be seen on every continent."[17] Collectively the evidence strongly indicates that humans are very much a part of nature and not in some way exempt from nature's rules.

That the concerns expressed by the environmental thinkers are shared by many people today can be seen in the large growth of environmental groups like the Environmental Defense Fund, the Natural Resources Defense Council, the Sierra Club, the National Audubon Society, the National Wildlife Federation, the Wilderness Society, Friends of the Earth, and Greenpeace. These organizations have millions of members, whose awareness of threats to the environment is supported by ever more sophisticated scientific research and motivated by a concern for their children and future generations. A new public view of nature is emerging that recognizes the complex integrated relationships that bind all living things together. We are becoming increasingly aware of the cumulative and potentially disastrous impact of imprudent use of resources on ecological systems and on all forms of life including humans.

Walter and Nancy Stewart endured some ridicule by allowing their lawn to become a meadow. They made this change for reasons that involved a sense of beauty, but they were also influenced by the *new* sense of nature handed to us by science, writings, cultural events, and the evolution of that phenomenon we call environmentalism: the sense that we are not independent of nature but reliant upon it for our continued existence and that humanity can have a profound influence on the course that nature takes on this planet. The Blums are consciously being "environmentalists" when they encourage native plants and animals to grow and live in their yard. It seems likely that Michael Pollan and Joel Meisel were also influenced by a new sense of nature when they decided to dissent from the conventional lush green lawn.

This new understanding of nature, our role in it, and our ability to influence its course has Americans thinking in new ways. Many of us are desperately concerned about the global environmental problems that face us. We wonder, just as desperately, what we can do to help alleviate these problems that often seem so vast they leave us feeling helpless and powerless. However there are many things we can do to tackle serious global environmental problems. Some opportunities for change lie just beyond our front and back doors: the lawn. By careful thought we can express our concern for nature by adopting an ecological approach to designing and caring for our lawn. We can strive to preserve some of the lawn's cherished values while diminishing some of its contributions to local, regional, and global environmental crises. For many people, their lawns provide a clear opportunity to "think globally, but act locally."[18]

Chapter 3
The Economic Juggernaut

Making the decision to have something other than a pure grass, continuously green, manicured lawn involves bucking a universally accepted view of what a yard should contain. In sports parlance, it is Westwood High School against the Chicago Bears. Here we explore the idea that this unequal match is a result of the powerful economic forces in a market economy. Let's begin by asking, what exactly is a lawn?

The Freedom Lawn

Lawn, n. a stretch of grass-covered land, esp. one closely mowed, as near a house or in a park.

Broadly interpreted, this definition does not prohibit the presence of other plants. In practice, it is easy to find mowed lawns that include many kinds of plants other than grass plants. A few common to northeastern lawns are: dandelion, violets, bluets, spurrey, chickweed, chrysanthemum, brown-eyed Susan, partridge berry, Canada mayflower, various clovers, plantains, evening primrose, rushes, and wood rush, as well as grasses not usually associated with the well-manicured lawn, such as broomsedge, sweet vernal grass, timothy, quack grass, oat grass, crabgrass, and foxtail grass. All of these plants can coexist quite nicely with grasses considered to be *the* lawn grasses, such as bluegrass, ryegrass, and fescue. All of these potential inhabitants can tolerate mowing since they can keep sufficient energy-fixing apparatus below the level of the mowing blade. Some will remain green and healthy but never produce flowers since their erect flowering stalks are cut off by the mower.

We call this the "Freedom Lawn" for it permits all kinds of plants to exist in the only way they know how—by growing (figure 18). The Freedom Lawn results from an interaction of naturally occurring processes and the selective effects of lawn mowing.

The Freedom Lawn is continually bombarded by seeds from nearby herbs, shrubs, and trees. Some of these may find an open space and germinate producing a new plant which, if it can tolerate the whirring blade, will become part of the lawn. One of the most interesting things about the process of plant establishment is that it does not occur equally everywhere. In most lawns there are subtle variations: moist depressions, droughty tops of mounds, somewhat cooler north-facing slopes or warmer and drier south-facing slopes, some areas always exposed to full sun and some partially shaded by nearby trees. These little variations in location alter the conditions of growth and survival of plants; some plants are better adapted to particular conditions than others. Thus the flora is far from uniform; the plants that compose the lawn tend to vary in response to the microenvironment and competition from other plants, that is, moist spots will have an abundance of species that thrive in moist environments, whereas species that best compete for scarce water will be found in droughty spots. Thus, the plant arrangement in the Freedom Lawn is well designed by the interaction between mowing and local ecology. The plants that succeed here do so without artificial intervention and collectively produce a green cover that is

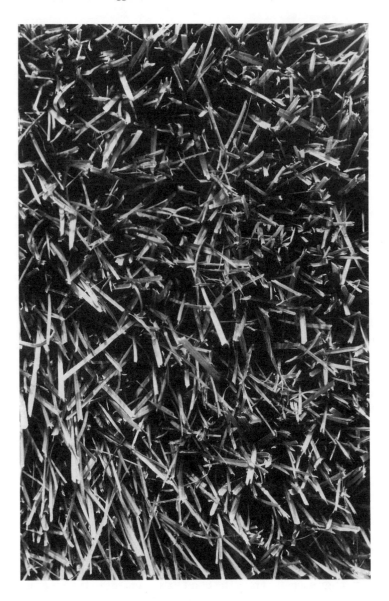

adapted to the peculiarities of place. In other words, with relatively little effort it is possible to have a green lawn adapted to the site.

Figure 18. The Industrial Lawn (on left) and the Freedom Lawn. Photographs by Lisa Vernegaard.

An important part of the adaptability of the Freedom Lawn is its built-in capacity to tolerate various kinds of stress. For example, because of the diversity of plants, a particular disease or insect is unlikely to wipe out the lawn. Some plants might suffer, but others would prosper and expand into areas vacated by the affected plants. The same would be true during prolonged droughts; the lawn might turn brown, but it is unlikely that it would die. Many plants have special mechanisms for surviving drought so they are able to resume active growth and return greenness to the lawn when the moisture returns.

The Industrial Lawn

In the sweeping parklike arcades of suburbia with continuous lawns following the graceful curves of residential roads, the Freedom Lawn is seldom found. "Freedom" in the lawn violates possibly all of the cardinal principles underlying the acceptable suburban lawn—principles that any hardware store or lawn care center shouts at you by means of their lawn care products. These products include grass seed specially designed for sun, shade, or in between; boxes and sacks of fertilizer of different formulations; squares of grass sod for instant lawns; bottles, cans, and sacks of insecticides and herbicides; hoses, sprinklers, and irrigation control devices; edgers and spreaders; mulching and sweeping machines; aerators and rollers; how-to books of lawn care; and mowers ranging from the inexpensive push mower hidden in a corner to the cadillac four-wheel drive rider mower featured on a pedestal. All of these support a different definition of the lawn.

Lawn, n. a stretch of grass-covered land especially that near a house or in a park, that is regularly and closely mowed, continuously green and to the greatest possible degree, free of weeds and pests.

This is the lawn your neighbor expects to see on his trip home after battling the evening rush hour; the lawn created by what has come to be known as the lawn care industry. In fact, we might think of it as the "Industrial Lawn" (figure 19).

Figure 19. An all-grass, pest-free, continuously green, and constantly mowed Industrial Lawn. Photograph by Peter M. Miller/Image Bank.

The Industrial Lawn rests on four basic principles of design and management: (1) It is composed of grass species only; (2) it is free of weeds and other pests; (3) in as far as possible it is continuously green; (4) it is regularly mowed to a low, even height. In contrast to the Freedom Lawn, which in the hands of a relaxed lawn owner requires little other than an occasional mowing, the Industrial Lawn is utterly dependent on the expenditure of considerable money, time, and energy. The Industrial Lawn is not attuned to the peculiarities of place. Like energy-intensive agriculture it ignores microclimates and species diversity and substitutes technology for natural processes. It has the added virtue, at least in terms of the lawn care industry, of never being completely attainable. There is always some new and necessary bit of technology, some new finding on fertilizers, some modification of pesticides or new variety of grass required to move toward the ideal or, in more competitive terms, to keep up with the neighbors.

How did such a powerful paradigm arise? The existence of an estimated $25 billion a year turfgrass industry[1] might raise suspicions of a smoke-filled room with CEOs plotting the manipulation of American homeowners and rubbing their hands together in anticipation of the increasing flood of money as the message of technological dependence is spread by wily lawn care promoters. Of course such is not the case. Instead the Industrial Lawn seems to have developed from a marriage of many forces: the long history of our love of the lawn coupled with technological advancements that made modern agriculture possible, agricultural corporations searching for new outlets for their products, and skilled marketeers fighting for market share.

Decades ago entrepreneurs identified the lawn as a potential market. They adapted existing agricultural technology to lawn management and design and through research developed new technology. They participated in the development of private and publicly supported organizations that promoted the care and culture of turf. Using marketing skills and advertising they were able to sell their products and gradually shape the concept of the lawn to meet their desire for increased profit. Thus arose the Industrial Lawn, by some standards a good thing, for it underlay a diverse industry that created jobs and income and contributed to the gross national product.

The American lawn in its industrial form presents a powerful symbol that explains, in part, the resistance that Meisel, the Stewarts, and the Blums met from their neighbors when they decided to change their lawns. Manicured green lawns unite the front yards of millions of suburban houses on both sides of the street to create an expansive and unifying parklike aspect. Michael Pollan captures the primitive force of this deeply ingrained view of the lawn: "To stand in the way of such a powerful current is not easily done. Since we have traditionally eschewed fences and hedges in America, the suburban vista can be marred by the negligence—or dissent—of a single property owner. This is why lawn care is regarded as such an important civic responsibility in the suburbs, and why, as I learned as a child, the majority will not tolerate the laggard. . . . That subtle yet unmistakable frontier, where the crew-cut lawn rubs up against a shaggy one, is a scar on the face of suburbia, . . . an intolerable hint of trouble in paradise."[2]

Growth of the Lawn Care Market

Like the crown of a king, any patch of lawn represents riches imported from the far corners of the kingdom. Fertilizers, pesticides, irrigation systems, water, tools, gas and oil, labor, and the very seed itself are all convened for the purpose of growing greener grass for a longer part of the year. Lawn and lawn care demands have created a large and steady market within each of these related industries. It is the scale and arrangement of these different components that describe the industry's regional characteristics, the variety of distinct lawn markets and overall changes in the direction of the industry.

One hundred years ago most Americans would have laughed at the idea of a lawn industry. Pastures were common and special measures to grow grass domestically would have seemed unusual. Today the total area in turfgrass is about 25 million acres[3] or 40 thousand square miles, an area slightly less than that of Pennsylvania. Grass is the selected ground cover for athletic fields, parks, playgrounds, highway verges, cemeteries, golf courses, schools, and homes. Eighty-one percent of the total, over 20,000,000 acres, is in home lawns.[4]

How did a large and complex industry evolve out of an individual's pursuit of a greener lawn? The lawn industry got its start in 1901 when the U.S. Congress allotted $17,000 to study the "best native and foreign grass species . . . for turfing lawns and pleasure grounds."[5] In 1920, the industry gained real momentum when the Greens Section of the United States Golf Association lobbied and won the support of the U.S. Department of Agriculture for a program researching grass species suitable for greens and fairways. Today grass research centers are found in most states, and courses on turf and its maintenance are offered by agricultural universities. The private and public infrastructure promoting the lawn and its maintenance is huge. The lawn industry has become part science, part aesthetics, part peer pressure, and increasingly part marketing. Homeowners are caught in a Sisyphean cycle of seeking a greener and greener lawn, a cycle curbed only by their willingness to pay.

In marketing, perception is everything. In a 1981 study of 800 lawn owners in the Piedmont and Coastal Plain regions of Virginia, 80 percent thought their lawn was average or below and were not satisfied with its present condition.[6] There will always be a lawn that is greener than yours, be it on a golf-course, in a photograph, or covering a neighbor's front yard. Thus, the stage is set for an infinitely expanding array of lawn care products and accessories.

The lawn care industry has become a growth industry constantly searching for new rationales for the expansion of lawns or the expansion of sales of lawn care products. Two important organizations, the Professional Lawn Care Association of America (PLCAA) and the Lawn Institute, both of which promote lawn and turf, now advertise not just improved property values resulting from greener lawns but increased health benefits as well: "Healthy turf means healthy lives." The PLCAA's and the Lawn Institute's brochure, "ABCs of Lawn and Turf Benefits," includes groundwater enhancement, buffering sports injuries, helping to discourage littering, providing a link with nature, and being generally therapeutic for humans.[7]

The Lawn Institute and the PLCAA state that a 50-by-50-foot lawn produces enough oxygen to sustain a family of four, an assertion that chapter 4 will show to be inaccurate. They also lobbied heavily in support of the 1990 Congressional Farm Bill which advocated tree and lawn planting in urban areas to combat noise, dust, and global warming. Clearly these ecologically and socially sound ideas will also produce economic benefits for the lawn care industry: the more grass planted, the more lawn care products sold. Occasionally segments of the industry advocate changes favorable to a sounder ecology: In 1990 the industry launched a national campaign to promote grasscycling—"today's turf tomorrow's earth." Several months of press releases and other publicity advocated leaving the cut grass on the lawn after mowing to increase the return of nutrients to the soil and to reduce yard waste entering landfills.

Organizations such as the Lawn Institute are making some legitimate attempts to promote environmentally sound lawn care, but the lawn industry as a whole is also capitalizing on marketing strategies that appeal to Americans' new environmental consciousness. One publication, *Lawn News,* titled an article, "Sell the Environment to Increase Your Profits."[8]

Dimensions of the Industry

The word *grass* is believed to have come from the old Aryan word *ghra,* also the root of grain, green, and grow. Grass was elemental, the "general herbage" that sustained life. "All flesh is as grass. And all the glory of man is as the flower of the grass" (Isaiah 40:6). Today healthy, green, growing grass has become more of a symbol of success and plenty rather than the annual crop on which life depends. Grass grown for lawn is not a food, not a medicine. It has become for many a luxury item basic to being a satisfied and responsible homeowner.

Because it is a nonessential product, most of the documentation concerning the size and complexity of the lawn industry comes from organizations with a vested interest in the business. For example the *Annual State of the Industry Report,* prepared by the editors of *Landscape Management,* is based solely on the experience of their membership, a self-selected group of corporations. Professional magazines generally publish information of interest to their readership, covering topics like pest management, golf course design, sports turf, and the latest grass seed hybrids. The magazine titles call out to their readership: *American Lawn Applicator, The Greenmaster, Golf Course Management, Outdoor Power Equipment, Grounds Maintenance,* and *Turf News*. These magazines do not publish many concrete statistics, however. Data detailing national trends and regional markets are high-priced trade information and the very stuff of competition. The profits of these corporations and professional lawn care services are based on knowing the product and the market better than the competition.

Much of the available documentation pertaining to lawns derives from the National Gardening Association's annual survey conducted by the Gallup Organization.[9] This unaligned poll is the most objective description of the industry available.

The 1991–92 survey indicates that 58 million American households (62 percent of all U.S. households) are engaged in lawn care. Table 1 shows that lawn care exceeds all other gardening activities by a considerable margin. These lawns cover an esti-

Table 1
Household Participation in Lawn and Garden Activities.

Activity	Households Participating (in millions)
Lawn care	58
Flower gardening	39
Indoor houseplants	39
Insect control	33
Shrub care	30
Vegetable gardening	29
Tree care	26
Landscaping	25
Flower bulbs	24
Fruit trees	14
Container gardening	12
Raising transplants	11
Herb gardening	8
Ornamental gardening	7
Growing berries	7

Source: National Gardening Association, *National Gardening Survey, 1991–1992* (Burlington, Vt.: National Gardening Association, 1992), 11.
Note: The survey is based on personal interviews conducted at home with a representative sample of U.S. households in all fifty states. Total numbers exceed 92.8 million because many homeowners engage in more than one lawn and gardening activity.

mated 20 million acres with an average size of about one-third of an acre.

These home lawns, green witnesses to countless American Saturdays, collectively occupy more acreage than any agricultural crop and account for the lion's share of the lawn industry's profits. Lawns range from highly manicured "estate" lawns to relatively low-maintenance lawns; in essence from the Industrial Lawn to the Freedom Lawn. Lawns at the industrial end require considerable annual investment whereas those at the freedom end require only a minimum investment. Millions of lawn owners lie in the middle range, each managing their average one-third acre by fighting sporadic battles against poor soil fertility, drought, insects, and incursions of children and wildlife. Although low-maintenance lawns constitute a considerable acreage, major profits for the lawn care industry come from upper and middle levels.

In 1991, the National Gardening Association estimated annual retail sales of residential lawn care products and equipment to be $6.9 billion. The Lawn Care Institute estimates that annual turf and lawn maintenance is altogether a $25 billion industry.

One might wonder how lawn care expenditures compare to expenditures made to raise agricultural crops. Although the comparison is a complex one, Goldin has reported that the Industrial Lawn often costs more per acre to maintain than it does to raise a crop of corn, rice, or sugarcane.[10]

In the context of one's own backyard, $25 billion is a difficult number to fathom. The lawn industry represents only one-half of

Grass for Happier Living

1 percent of the United States gross national product, but this percentage in no way adequately reflects the industry's power to determine the structure of our landscapes and to influence our psyches and our attitudes toward our environment.

Components of the Industry

The components of the lawn industry are determined by the four principles underlying the Industrial Lawn: all grass, weed and pest free, continuously green, and closely mowed. To establish a lawn in the first place, one requires a source of grass seed. After it has been sown, a grass plant uses the sun's energy to produce plant biomass: leaves, roots, stem, flowers, and fruits. The amount of biomass is determined in large measure by the amount of water and nutrients such as nitrogen, phosphorus, and potassium present and available in the soil. The grass plant's habit of growing up from below, probably an evolutionary adaptation to grazing animals, makes it ideal for mowing but also creates a recurrent need for mowing, raking, and various other maintenance tasks. To encourage a lawn to be continuously green, fertilizers (nutrients) and water may be added. Some insects and fungi view a lush green lawn as the centerpiece of a huge Bacchanalian feast. Thwarting the desires of these pests requires the application of insecticides and fungicides. To inhibit the growth of weeds, herbicides can be applied. The endless nature of this cycle is clearly evident: water and nutrients promote growth; herbicides lessen competition from nongrass weeds; rapid growth requires frequent mowing; pesticides inhibit not only grass predators but also organisms that decompose clippings; to achieve the desired appearance, clippings require removal; the removal of clippings requires fertilizers to replace lost nutrients. The cycle is music to the ears of the lawn care industry.

Seed Seed usually accounts for only 2 to 3 percent of the annual costs of maintaining a home lawn, but that small investment for each lawn adds up to an important segment of the billion dollar seed industry. Up until World War II, grass seed was traded as a commodity like most other agricultural seeds. The market was full of different varieties of grass supplied by many individual growers and sellers. As the seed market has become increasingly industrialized, producers have had to look for new ways to hold on to their market share. In recent years, producers have transformed grass seed into a specialty product in an effort to appear different and better to the perplexed consumer.

Perusal of the packages of grass seed at your hardware store will give an idea of current approaches to marketing problems. For example, producers now advertise seed mixtures for heavy traffic, low maintenance, and deep shade. An alternative that seed companies might explore is the development and promotion of a more species-diverse lawn—what we are calling the Freedom Lawn. The value of such an alternative will be discussed in our concluding chapter.

Fertilizer Fertilizer, like other components of lawn care, is a product originally developed for agriculture. As agricultural markets stabilize or stagnate, the fertilizer industry seeks growth through expansion into the lawn care market. By weight only 5 to 10 percent of the fertilizer sold in the United States is purchased to fertilize lawns, but this market accounts for 25 percent of the industry's profits. In fact there is little difference between lawn fertilizers and agricultural fertilizers other than the size of the package and the type of marketing.[11]

The marketing strategy adopted by the fertilizer industry involves convincing the homeowner of the need for repeated applications of fertilizer to the lawn. The concept of the continuously green lawn and the removal of grass clippings from the lawn contribute to this need. Removing grass clippings from the lawn is analogous to removing the ears of corn or silage in corn culture; fertilizers become necessary to make up for nutrients lost through removal.

The $17 billion fertilizer industry (connected to both agriculture and lawn care) is highly dependent on mineral resources and abundant energy supplies.[12] Fertilizers for lawns are a mix of nitrogen, phosphorus, and potassium, the familiar N-P-K, just as agricultural fertilizers are. The production process is entirely dependent on fossil fuels, particularly oil and natural gas. Natural gas is the main ingredient in the production of nitrogen fertilizers and the main reason why oil companies with natural gas holdings own fertilizer companies. Although fossil fuels are not a direct component of the production of phosphorus and potassium, they are heavily used in mining, processing, and transporting these nutrients to the market.

Growing public concern over the environmental impact of heavy fertilization is exerting some pressure on the fertilizer industry. Some fertilizer companies are responding to this pressure by developing "natural" fertilizers. The Ringer Corporation manufactures some of its fertilizers from organic by-products of other industries. Nitrogen is derived from feather meal from the poultry business, phosphorus from ground bone, and potassium from sunflower seed ash.[13] For the environment, this is clearly a step in the right direction, for at one fell swoop the consumption of fossil fuels is reduced, and waste products that might otherwise contribute to our solid waste problems are put to profitable use.

Pesticides The world market for pesticides reached nearly $25 billion in 1991.[14] Sales of lawn care pesticides in the United States make up a surprisingly large part of the total world expenditure. In 1988, more than 700 million dollars worth of pesticides—about 67 million pounds of chemicals—were sold for use on American lawns.[15] To understand the marketing strategy of the chemical lawn care companies, one need only look at the bewildering array of pesticides present in any hardware store during the summer months (see figure 20).

One of the principal marketing strategies of the pesticide industry focuses on the chemical destruction of pests, organisms that compete with, destroy, or devalue agricultural products, including lawn grass. Advertising often projects a rather neat world where technology simplifies the control of weeds, insects, and plant disease and produces an ordered world under human control.

In the 1970s many citizens began to fall out of love with synthetic chemicals, including poisons such as those used in many turf pesticides. Chemical pesticides used in lawn care are designed to destroy or control living organisms and may therefore pose a threat to many forms of life including humans and their

Figure 20. Hardware stores stock a wide array of chemicals designed to aid a homeowner with every imaginable lawn maintenance problem. Photograph by G. Carleton Ray.

pets. Fewer than 1 percent of the estimated 500,000 species of plants, animals, and microbes in the United States are considered pests.[16] The other 99 percent carry out an array of essential functions such as decomposing organic wastes, degrading pollutants, recycling nutrients, moderating the structure of the soil, preserving genetic diversity, and serving as vital parts of food chains.

Many of these organisms live in the soil. What is soil? From the point of view of an ecologist or agriculturist, soil is a complex substance, the product of many years of occupancy by plants, animals, and microbes. Soil is composed of minerals, sand, silt, and clay; it contains thousands of organisms whose physical and biochemical actions make soil a dynamic body. These organisms decompose organic matter, renewing the store of soil nutrients. Soil also retains water, and this capacity, together with the presence of nutrients, enables plants to take root and grow. Usually, the soil is only one to two feet deep with most activity and nutrients concentrated in the upper six inches or so—the topsoil.

Beneficial organisms, including organisms in the soil, can become the inadvertent targets of pesticides applied to kill known pests. In the 1960s Rachel Carson voiced concern about unintended targets and human health, and since then a stream of investigators have echoed her apprehension.

As a result, chemical companies have found their pesticide markets influenced by mounting public concern. In a survey conducted by *Lawn Servicing* magazine, 64 percent of lawn care professionals said customers confronted them with questions about pesticides. Thirty-three percent had cancellations as a result of what they called "bad publicity."[17]

Irrigation   Grass needs water to grow (figure 21). The amount of water that a yard demands (or its owner thinks it demands) is subject to variables in addition to place and climate. The type of landscaping, the varieties of grass grown, the type of soil, the

Figure 21. The Industrial Lawn often requires watering to preserve an ever-green appearance. © Photograph by Karen Bussolini, 1992.

pressure from neighbors, and the degree of importance the owner places on having a green lawn are all factors contributing to the amount of water applied.

In arid regions of our country, such as the Southwest, irrigation is absolutely essential to maintain a lawn: relatively large quantities of irrigation water are required to maintain a lawn under the high rates of evaporation common to these areas. Without irrigation the lawn would soon be replaced by local vegetation adapted to frequent drought conditions. In the more humid regions of our country, many lawns survive quite nicely without supplementary water, although they may become dormant and turn brown during periods of low rainfall.

The lawn irrigation market is built on technology designed to supply water. Delivery systems range from sprinkler cans, through hoses with attached sprinklers, to elaborate underground sprinkler systems. Attachments include fairly sophisticated technology like timers and soil-water-sensing elements that automatically turn sprinkler systems on and off. One thing that all of these devices have in common is the need for a source of water. Thus two kinds of profit are involved in the irrigation market: profit from the sale of technology and profit from the sale of water.

The National Xeriscape Council,[18] an organization that promotes landscaping with a minimum application of water, estimates that up to 30 percent of urban water on the East Coast is used for lawn irrigation. In the West they estimate that 60 percent of urban water use is for lawn irrigation.[19]

Not all eastern lawns are heavily watered. The Turfgrass Council of North Carolina found in that state that lawn watering is low on most homeowners' list of priorities and that more than 75 percent of homeowners never water their lawns at all.[20]

As fresh water becomes scarcer and an increasingly valuable resource, many municipalities are taking a harder look at regulating lower priority usage. The recent combination of prolonged drought and recession in Southern California may lead to the regulation of lawn watering by city ordinance. In 1990, Santa Barbara, California, banned lawn watering all together, forcing resourceful homeowners to consider other options such as buying water from private companies, recycling "grey" water from the house, or using vegetable dyes to make their lawns green. Some homeowners choose to go to great lengths to keep their yards looking like oases in the desert! Even regions considered to have ample water supplies have had to regulate water use for lawns and gardens during recent dry periods. The notoriously wet city of Seattle, Washington, required homeowners in 1989 to conserve dwindling water supplies by watering their yards only every third day. Neighbors eyed each other cautiously wondering if anyone would notice infringements of the new laws. Telltale green lawns exposed those who ignored the ban. Eventually the rains came, but the drought had triggered an interesting twist on the standard peer pressure surrounding the ideal of the continuously green lawn.

Lawn Equipment Mowing is a requirement for any lawn, but manicuring is an essential requirement of the Industrial Lawn. The lawn equipment industry thrives on sales of tools designed to achieve a perfectly groomed lawn (figure 22).

A special tool exists for every conceivable lawn maintenance task: these include edgers, weed eaters, fertilizer spreaders, dusting and spraying equipment, leaf blowers, precision seeders, turf aerators, rotary and reel mowers, and rider mowers. In 1991, U.S. industry shipped an estimated $4.6 billion in lawn and garden equipment to both domestic and international markets, primarily Canada and the European Community.[21]

The equipment market has traditionally depended on sales of new improved state-of-the-art gadgetry. Currently the most popular state-of-the-art gadget is the large, expensive, and profitable rider mower. In 1989 rider mowers accounted for fewer than

Figure 22. Lawn care can involve simple hand tools, but today's industry has promoted a vast array of motorized labor-saving tools. © Photograph by Karen Bussolini, 1992.

20 percent of all mowers produced but more than 50 percent of total sales value.[22] The demand for rider mowers is only partly linked to new housing construction; profits can also be generated by consumers who "step up" from their old walk-behind power mowers to the "convenience and comfort" of new rider mowers.

Like the fertilizer and pesticide markets, the mower markets are feeling the impact of ecological thinking. A rash of new mowers designed to leave the grass clippings on the lawn has arrived at local hardware stores. The environmental rationale is that the clippings fertilize the lawn when they decompose and release nutrients to the grass roots and that by leaving clippings on the lawn the stream of yard waste going to the local landfill is reduced. This is bad news for the fertilizer industries since it reduces the need for fertilizers.

Sod To install an instant lawn, homeowners can purchase sod, grass grown as an agricultural crop (figure 23). Sod can be used

Figure 23. The production of sod requires large inputs of water, fertilizer, pesticides, and fossil fuel. Photograph by Grant Heilman/Grant Heilman Photography, Inc.

to establish replacement lawns, but the primary market for sod is new houses with unlandscaped yards. The price of sod is an important start-up cost associated with a new home for homeowners who want a lawn and want it now. As the housing market goes, so goes the market for sod. Housing starts have dropped from about 1.8 million in the mid-1980s to less than 1 million in 1991. This decline was closely shadowed in the sod market.[23]

Sod is an expensive proposition. In 1991, wholesale sod could be purchased in lots of 506 square feet for $150. For a modest lawn of 5,000 square feet this would entail an investment of about $1,500, not including the inevitable retail markup and installation fee. Do-it-yourself homeowners could purchase sod from garden supply stores in units of 9 square feet for $3.50. At this price, a 5,000 square foot lawn would cost about $1,900 not including other costs such as delivery, site preparation, fertilizers, and so forth.

A principal environmental question associated with sod farming is its effect on the agricultural soil supporting the sod farm. Since an amount of soil is removed with each sod harvest, sod farming would seem to be a soil-mining process that would lead to the destruction of the original soil. At a time when there is great concern about the erosion and destruction of agricultural soil worldwide, it might seem a questionable practice to destroy agriculturally productive soils to produce instant lawns. However, many new lawns are constructed using topsoil imported from some other place. That place might well be a former farm or a forest—so what is the difference between mining only topsoil and mining sod with topsoil?

Fortunately, it is possible to make new soil! Most of the exposed earth we see around construction sites or other disturbed places is not soil as defined earlier, but rather raw earth from which soil is made by ecological processes. The copious addition of organic matter can often turn these raw earth materials into reasonably decent soils. In our society today, there are two plentiful sources of organic matter, sewage sludge and leaves collected from street trees. With appropriate attention from government, industry, and individuals, these two major sources of solid waste could be put to profitable use transforming raw earth into fertile soil.

Labor   All grass has one consistent characteristic, it grows taller and taller. Keeping it short is the lawn care industry's bread and butter. This industry was built on manual labor, from early teams of European scythers to summer jobs for American teenagers. In today's market, labor remains the largest single cost in both home and commercial lawn care.

Labor is what makes the lawn care industry an important addition to a local economy and might be used in an argument justifying intensive lawn care operations. However, as the main expense in the industry, it is the first area targeted for cutbacks in a competitive economy. The lawn situation might be somewhat analogous to the intense argument raging in the Pacific Northwest about cutting old-growth Douglas fir forests. Industry maintains that reducing cuts will cost the loss of jobs, but at the same time industry is furiously eliminating jobs through the introduction of labor-saving technology.

Homeowners' Expenditures

To determine how much the "average" homeowner spends is a daunting task. Several states (Maryland, 1987; New Jersey, 1983; New York, 1977; North Carolina, 1986; Ohio, 1989; and Pennsylvania, 1991) have conducted surveys, but due to different methodologies and assumptions it is hard to compile "average" numbers from their results. Nevertheless, some generalizations are still possible.

The bulk of turfgrass covers home lawns. In New Jersey, home lawns represented three-quarters of the total lawn acreage in the state and half the annual maintenance costs. Ohio home-owners accounted for nearly two-thirds of the state's lawn acre-age, and they spent 64 cents out of every dollar spent on lawn care. About two-thirds of the lawn acreage in North Carolina is in home lawns, and homeowners accounted for one-half of the statewide expenditures.

Because the average size of the home lawn is only one-third of an acre, the expenses incurred in lawn care are quite high on a per acre basis, much higher than the per acre cost for agri-cultural crops, for example. The average household has a mower and a variety of other tools and makes annual purchases of new tools, fertilizers, and pesticides. To gain some idea of the magni-tude of these expenses we will look at the North Carolina Turf-grass Survey. North Carolina is a state of numerous small cities with an extensive rural area and may not be representative of the nation as a whole. Nevertheless the statistics from North Carolina can give us some idea of how we spend our money on lawn maintenance.

Table 2 shows the total amount of money that North Carolina homeowners spend on lawn care. The average lawn owner and industry's ideal lawn owner are two different people. The indus-try ideal is a figure of lore, industry news releases, and do-it-yourself gardening guides. The North Carolina Survey, based on personal interviews, yields a more realistic picture. North Caro-lina's average home lawn is 0.6 acres, twice as large as the na-tional average. Homeowners spent $407 on their lawns each year: $158 was for maintenance while the remaining $249 went to purchasing new equipment. The lawn is many things to many people, but even so it represents an annual expense of hundreds of dollars for most households and likely much more for higher-income homes.

Table 2
Lawn Maintenance Expenses for North Carolina.

Practice	Cost (in thousands of dollars)	Percentage
Labor: Mowing	$143,902	42.5
Mowing equipment and gas	113,768	33.6
Repairs	23,363	6.9
Fertilizer	22,686	6.7
Seed	9,480	2.8
Irrigation equipment	5,418	1.6
Weed control products	5,079	1.5
Lime	3,724	1.1
Insect control products	3,386	1.0
Other	7,788	2.3
Total	$338,594	100.0

Source: Turfgrass Council of North Carolina, *North Carolina Turfgrass Survey* (Raleigh, N.C.: North Carolina Crop and Livestock Reporting Service, 1987), 35.

Professional Lawn Care

As Americans cope with an economy that requires longer work hours and two wage earners, mowing the lawn can lose its attraction. Lawn care services provide an alternative that claims to cost no more than a homeowners do-it-yourself investment of money and time, but with more professional results. Professional lawn care services weave all of the lawn industry components into "greenscapes"—the only thing left to the lawn owner is writing the check.

The lawn care service industry has experienced unprecedented growth in the last decade and demand is still increasing. Today there are approximately 5,500 professional lawn care companies in the United States. These range in size from one person with a pickup truck to major public corporations with franchises across the country. Nine and a half million American homeowners hire professional lawn care service. These services tend approximately 1.6 million acres of residential lawns and 0.5 million acres of institutional and commercial lawns.[24]

A Washington-based consumer advocacy magazine solicited bids for professional lawn services for five different lawns in the Washington, D.C., metropolitan area. The survey showed that companies often recommend different treatment, and thus bids may vary widely. For one lawn, eight companies provided bids ranging from $180 per year to more than $3,000 for a total make-over.[25]

In January 1989 the professional lawn care industry was grossing $2.8 billion.[26] The key to profit in the lawn care industry is to sell services. Services include mowing, fertilizing, insect and disease control, aeration, seeding, dethatching, sodding, renovation, edging, and shrub and small tree care. Financial success rides or falls on persuasive salesmanship. Companies offer a variety of packages that differ most in frequency of chemical application and attention paid to surrounding plants. Often they offer a lawn owner a menu of lawn care more complicated than the dinner menu at the Waldorf Astoria. Not infrequently they have been the target of heavy criticism for careless or deceptive advertising about the efficacy and safety of pesticides used in the lawn care procedures they advocate. In 1990, the U.S. General Accounting Office (GAO) issued a report about the claims that lawn care companies were making about the products they use. Staff from the GAO identified themselves as private citizens, and called several lawn care companies to ask about product safety. Although some companies acknowledged that there were environmental and health risks, several others claimed that their products were safe or nontoxic. Chapter 4 describes several health-related problems that contradict these safety claims. Responses that the GAO found to be misleading or false included:[27]

"The only way to be affected by [the pesticide] 2,4-D would be to lay [*sic*] in it for a few days."

"The safety issue has been blown out of proportion. Such a small amount of chemicals are put down directly on plants . . . . [They do] not affect animals or people."

"All chemicals [used] are nontoxic."

"Dogs may get a rash or irritated [from diazinon], but they will only feel a little itchy. This is the same reaction the applicator gets when the pesticide touches their [*sic*] skin."

The lawn care industry thrives on and is probably the chief promoter of the Industrial Lawn. The idea of the all-grass, pest-free, continuously green, and frequently mowed lawn is money in the bank.

The Clash of Philosophies

The Freedom Lawn and the Industrial Lawn strategies embrace extremely different philosophical views of nature. Although a product of human management, the Freedom Lawn has a large element of what Wes Jackson has called "nature's wisdom." The pattern of the Freedom Lawn is in large measure the result of natural processes. Most of all the Freedom Lawn emphasizes the use of solar energy and minimizes the use of fossil energy and other scarce natural resources.

The Industrial Lawn, on the other hand, is under stricter human management. With the help of technology, humans try to control the pattern of biological relationships within the lawn. Plant diversity is minimized; grass predation by insects is controlled by insecticides; weed species are controlled by herbicides; fungal attacks on grass are thwarted by fungicides; the addition of fertilizers substitutes for naturally occurring nutrient cycling; droughts are avoided by irrigation; and mechanical soil aeration compensates for the absence of a soil structure that promotes natural aeration.

Relative to the Freedom Lawn, the Industrial Lawn depends upon fossil fuel energy, irrigation water, pesticides, and fertilizers. It requires a considerably greater drain on world resources. From an ecological point of view, the Industrial Lawn also causes unanticipated environmental side effects. These matters will be the subject of the next chapter.

Chapter 4
Environmental Costs

As you wheel out the lawn mower for the first time in the spring to subdue rampantly growing grass and defiant stands of dandelions, it is difficult to imagine that mowing your lawn might affect anything more than those dandelions, your peace of mind, and your neighbor whose view from the dining room window includes your yard. Yet, creating something as beautiful and wholesome-looking as a lawn can adversely affect the earth. The more we work to make the lawn beautiful by watering, spraying, and mowing, the more we remove it from the natural ecosystem that would exist if we had not interfered. What is the ecological price of making and maintaining a lawn?

It once seemed inconceivable that humans could affect the processes that maintain life on earth. The earth appeared simply too vast. With the Industrial Revolution, the growth of technology, the enormous growth in our use of renewable and non-renewable resources, and the explosion in numbers of human beings on earth, we are finding, as William Vogt suggested in the 1950s, that the earth has shrunk, that space is no longer the protection we once thought it was, and that indeed we are capable of affecting global ecological processes.

Global ecological cycles are being altered in ways inimical to human welfare. Many regional air and water pollution problems have developed. We are concerned about the productivity of the lands and the seas. So grave are these problems that we are daily bombarded with news about global warming, the ozone hole, and the disappearance of tropical rain forests. Indeed, many are concerned about the future of humanity. Innumerable conferences, such as the 1992 Earth Summit in Rio de Janeiro, discuss the dimensions of and the projected solutions to an ever-lengthening list of environmental problems. This is the context within which we wish to consider the ecology of the lawn, our own little piece of the biosphere.

Natural Ecosystems versus Human-Modified Ecosystems

The people, grass plants, earthworms, and other organisms that live in or pass through our lawns form a dynamic community or ecosystem where species interact with each other and with their physical and chemical surroundings. We sometimes forget that our lush green carpets can be viewed as such a system—that ecology is taking place in our own backyards. How does the ecology of the lawn compare to the ecology of more "natural" systems, and how do activities that take place in our yards affect the local, regional, and world environments in which we live? Our third of an acre may not seem like much, but lawns combine to cover 20 million acres of the United States, making grass the largest single "crop" and a significant part of the American landscape.

Although human civilization today affects even the most remote ecosystems from the tropics to the Antarctic, we can identify ecosystems in which humans exert relatively little day-to-day influence. These "natural" ecosystems are powered almost exclusively by the sun, which provides the energy needed by all the plants and animals and drives the water cycle and the circulation of nutrients within the ecosystem. Natural ecosystems, like a forest or a prairie, do not require external management because the organisms are adapted to the physical and chemical conditions of the site and the circulation of nutrients between the living and the nonliving is fairly complete. These ecosystems are solar powered, self-regulating and self-fertilizing (figure 24).

The vast area of the United States contains a wide variety of climates and soils. In presettlement times, natural ecosystems in Maine and many of the far western states were dominated by evergreen forests, large sections of the eastern states were covered with deciduous forests, the middle of our continent was clothed in grasslands, and large areas of the Southwest supported desert ecosystems. Today, only remnants of undisturbed, naturally occurring ecosystems remain. Agriculture, grazing, forestry, mining, and other land uses, which are essential to maintaining our way of life, have resulted in a substantial alteration in the ecosystems of presettlement times. However, the same climate and soil conditions that gave rise to the forests, grasslands, and deserts persist today, and if the influence of humans were removed, ecological theory tells us that after a time the regional patterns of naturally occurring ecosystems would reappear. For example, if we stopped maintaining our lawns in Tucson they would become minideserts, in Maryland they would start developing toward forests, and in Lincoln, Nebraska, the first stages of a prairie would appear.

Capability Brown's lawn was developed for the cool, moist, mild climate of England so favorable to growing grass. In America, we maintain lawns in areas that were once forests, grasslands, and deserts and do so under a variety of climatic conditions, many of which are not naturally favorable to turfgrasses. Lawns grown in these regions differ dramatically from the natural ecosystems they replace because grass grown in an environment where it would not naturally occur depends on human management and supplements of fossil energy, water, and chemicals for survival. The supplement required depends on the naturally occurring local climatic and soil conditions and on the kind of lawn we choose to maintain, with the Freedom Lawn at the low end and the Industrial Lawn at the high end in terms of supplements required. However, in the desert or semiarid climates, even the Freedom Lawn would not be possible without irrigation. In the desert, a yard would consist of native drought-resistant plants, stones, and patches of bare ground.

The use of fossil energy and chemical additives directly affects not only the ecosystem of the lawn but also interconnected ecosystems, such as air and water, and these supplements may have both direct and indirect effects on the health of many organisms including human beings. Using these supplements for the lawn may also limit the supplies of scarce resources that could be available for more vital uses.

Fossil Energy

Grass plants, through the process of photosynthesis, use the energy of the sun to grow. The lawn, particularly the Industrial Lawn, also requires the burning of carbon stored away in fossil fuels millions of years ago to power equipment and to manufacture and transport fertilizers and pesticides. The lawn is firmly linked to environmental issues that surround fossil fuel consumption.

During the oil crises of the mid-1970s energy was thrust into the spotlight as a national issue. Dramatic steps were taken to reduce the consumption of fossil fuels, including mileage standards for automobiles and tax breaks for energy-saving building improvements. Complacency about energy issues returned during the 1980s as oil prices dropped. The Persian Gulf crisis once again put energy consumption back on the national agenda as the hazards of depending on the oil-rich, but politically unstable Middle East again became apparent.

Political concern over fossil fuels has waxed and waned in response to supply, but scientific concern about the impact of the use of fossil fuel on our environment has steadily intensified. The

-Solar energy
-Precipitation
-Nutrients in rain and dust
-Atmospheric carbon
 dioxide
-Seeds from local region

INPUTS

NATURALLY OCCURRING FOREST OR GRASSLAND
STRUCTURALLY COMPLEX ECOSYSTEM

-Nutrients largely retained
-Net carbon accumulation
 in vegetation and soil

-(Relatively) large
 proportion of water
 evaporated to atmosphere
-Little surface runoff

OUTPUTS

-(Relatively) few nutrients
 lost in drainage water
-Less carbon dioxide output
 than input
-Very little soil erosion

CONSEQUENCES

-Much biological
 diversity: provides home
 for many species of
 plants, animals, and
 microbes

-Clean water to streams
 and ground water
-Reduces global warming
 by net removal of
 carbon dioxide from the
 atmosphere

-No impact on municipal
 land fills
-No impact on global
 fossil fuel supplies

-Solar energy
-Precipitation
-Nutrients in rain and dust
-Atmospheric carbon
 dioxide
-Seeds from local region

INPUTS

-Fossil fuel energy
-Irrigation water
-Nutrients in fertilizers
-Pesticides
-Grass seed or sod

AN INDUSTRIAL LAWN
STRUCTURALLY SIMPLE ECOSYSTEM

-Relatively fewer nutrients
 retained
-Net carbon loss when
 carbon in fossil fuels,
 directly or indirectly
 associated with lawn
 care, is counted

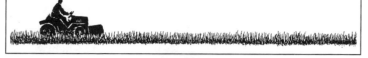

-More surface runoff
-More nutrients lost in
 drainage water
-Pesticides and fertilizer
 nutrients washed into
 neighboring water supply

OUTPUTS

-Carbon dioxide output
 greater than input
-Nutrients and pesticides
 removed in grass clippings

CONSEQUENCES

-Less biological
 diversity: local plant
 species displaced by turf
 grasses and turf-adapted
 animals and microbes **
-Contributes to increased
 global warming

-Increases stress on
 municipal water supplies *
-Increases municipal solid
 waste problems *
-Pesticides may con-
 taminate food chains *
-Pesticides on lawns may
 threaten human health *

-Disrupts biology of
 neighboring surface
 waters *
-Uses up global fossil fuel
 supplies

Figure 24. A general comparison between the relative environmental impact of a naturally occurring forest or grassland ecosystem and the Industrial Lawn that might replace it. Asterisks indicate consequences that would be minimized (*) or absent (**) if a Freedom Lawn were to replace an Industrial Lawn. The Freedom Lawn is mowed, but the grass clippings are not removed nor are fertilizers or pesticides applied.

use of fossil fuel has been increasingly linked to environmental problems including smog, acid rain, mega-oil spills, destruction of the ozone layer, and global warming. Energy problems do not just concern prices at the gas pump anymore, and environmental considerations such as global warming are not as far removed from decisions you make about the lawn as you might think.

In lawn management, the most obvious use of fossil fuel is in the mechanized equipment we use to groom our grass. Lawn mowers, aerators, leaf blowers, weed whackers, and edgers all consume fossil fuels, either directly (gasoline-powered equipment) or indirectly (electrically powered equipment). Approximately 13 million lawn utility machines are sold every year in the United States.[1] Most machines are powered by lightweight, two-cycle engines that use about one quart of gasoline per hour. The average homeowner spends about forty hours a year behind a power mower. Multiply this by 58 million homeowners and you get 580 million gallons of gasoline, not counting fuel for machines other than mowers. This represents only about 1 percent of the gasoline we use in cars, but, for a variety of reasons, this amount is more significant than this percentage might indicate.

Fossil fuels are also required to produce and ship fertilizers and pesticides. The principal nutrients in fertilizers are nitrogen (N), phosphorus (P), and potassium (K); these are listed on fertilizer bags as percentages by weight, for example NPK 10–10–10.

Although the air we breathe contains 78 percent nitrogen gas, plants cannot use nitrogen in this form. Some bacteria, such as those associated with the roots of clovers or alder bushes, can convert atmospheric nitrogen into a form usable by plants. In natural ecosystems, microorganisms that decompose dead organic matter also release nitrogen previously incorporated in plant or animal tissue in the process. Prior to World War I, most nitrogen fertilizers came from organic sources such as animal manure, guano, and bloodmeal. In 1913, Fritz Haber and Carl Bosch, two German scientists, learned how to capture nitrogen from the atmosphere by combining it with the hydrogen in natural gas to form synthetic ammonia, a nitrogen-rich compound. This discovery formed the basis for the synthetic fertilizer industry whose products require substantial fossil-fuel consumption and have increasingly replaced organic fertilizers.

Fossil fuels are needed to mine, refine, and transport potassium and phosphorus used in fertilizers. Each of these processes require the input of fossil fuels. The crude mined product has to go through refinements powered by fossil fuel before it is placed in a bag. The bag of 10–10–10 you buy in Macon, Georgia, may have originated, in part, in Florida, Utah, and Saudi Arabia; thus still more fossil fuel is required to power the ships, trains, trucks, and finally the car that will move these products from their source to your lawn. The manufacture and distribution of pesticides is similar to fertilizers in that fossil fuels are needed both as raw materials and in manufacturing and distributing these chemicals. Thus, evaluating the application of fertilizers and pesticides to one's lawn involves not only the direct effects of these substances but their hidden cost in terms of fossil fuel consumption.

As discussed in the previous chapter, the Lawn Institute and the PLCAA have stated that through photosynthesis a 50-by-50-foot lawn generates enough oxygen to meet the needs of a family of four. This kind of accounting fails to consider the oxygen consumed by microorganisms as they decompose grass clippings, nor does it account for the large amounts of oxygen consumed in burning fossil fuels associated with lawn mowing and the production and distribution of fertilizers and pesticides. When these factors are included in the calculations, the 50-by-50-foot lawn becomes responsible not for a gain but for a substantial net loss of oxygen from the atmosphere.

Although we do not often realize the energy consumption involved in watering our lawns, in arid climates such as the Southwest, irrigation can match, if not outstrip, the energy costs involved in lawn mowing. Water has to be pumped over various heights to move it from its source, such as the Rocky Mountains or an aquifer deep in the earth, to the place it is used in the dry Southwest, and pumping requires energy. The city of Irvine, California, estimates that watering one acre of lawn for one year consumes as much energy as mowing that lawn.[2]

The use of fossil energy is essential to our survival and well-being, but fossil fuels have serious and costly environmental and health effects: the burning of fossil fuels has been shown to contribute to the formation of acid rain, smog, ozone, and green-house gases and to cause respiratory health problems for residents in many large cities.

What is Global Warming?

Global warming results when the atmospheric content of greenhouse gases (carbon dioxide, methane, nitrogen oxides, and chlorofluorocarbons) is increased by human-generated pollution. These gases slow the loss of heat from the earth and warm the atmosphere (figure 25). Scientists are not absolutely certain that climatic change is actually occurring nor are they sure of how climatic change will affect the earth. However, they do know that "the average global temperature has already risen about 1 degree Fahrenheit in the last century, compared to a natural [change] of less than 1 degree per millennium over the last 12,000 years."[3] A degree or two may not seem like much, but it does not take a big change in temperature to cause big changes in our environment. During the last ice age, when much of the northern United States was covered with an ice sheet thousands of feet thick, the average global temperature was only 9 degrees Fahrenheit lower than today's. A warming of just 3 to 4 degrees over today's average temperature will make the earth warmer than it has been at any time in recorded human history. Many atmospheric scientists predict dire consequences from small average temperature changes, including sea level rise and major changes in weather patterns. The ice caps would melt, agricultural patterns would shift, low-lying coastal areas around the world would be flooded, and major shifts in the earth's vegetational patterns would occur. Human societies, which depend on the biotic stability of the globe, would be severely stressed and thoroughly disrupted by these changes.

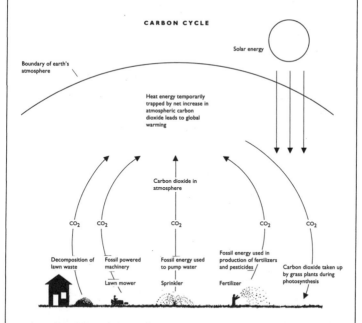

Figure 25. The carbon cycle as it pertains to the Industrial Lawn. As the diagram shows, the Industrial Lawn adds more carbon dioxide to the atmosphere than it removes, thus contributing to global warming.

Vehicles, power plants, factories, and space heating are primary causes of these problems, but the use of fossil fuels associated with lawn maintenance contributes to the problem. Indeed, the numbers are staggering. California's Air Resources Board has determined that in the one hour it takes to mow a lawn, the power lawn mower emits pollutants equivalent to driving 350 miles. Annual pollution emissions from lawn utility machines in California are equivalent to the emissions produced by 3.5 million 1991 model automobiles driven 16,000 miles each.[4]

Why are the emissions from lawn equipment so high? Lawn equipment generally uses small, two-cycle engines. Although light and mechanically simple, two-cycle engines produce significantly more pollutants than the more efficient four-cycle engine or automobile engines.[5] Indirect pollution generated by electrically powered lawn tools must also be considered. Electric motors may not release pollutants in your yard, but the power plant that produces the electricity releases pollutants elsewhere that contribute to health and environmental problems.

Although the emissions from an individual homeowner's mowing may seem small, the collective emissions generate an appreciable amount of air pollution. Recognizing the significant amount of pollution emitted by lawn care equipment, California passed regulations in 1990 that require manufacturers to reduce emissions by the year 2000. Influenced by California's regulations and the results of their own emissions study, the U.S. Environmental Protection Agency initiated development of small-engine emission standards for the whole of the United States in 1992.[6]

Chemicals

Although Rachel Carson opened our eyes to the health hazards stemming from chemical use in our environment, pesticide and fertilizer use on home lawns has steadily increased since the 1950s (see figure 26). According to the Environmental Protection

Figure 26. Many homes contain collections of partially used boxes, bottles, and bags of chemicals. © Photograph by Karen Bussolini, 1992.

Agency (EPA), a typical annual management program for an Industrial Lawn includes four or more applications of a high-nitrogen fertilizer and ten or more doses of various pesticides. The EPA estimated that in 1984 more synthetic fertilizers were applied annually to American lawns than the entire country of India applied to all its food crops.[7] Most Industrial Lawns receive between three and twenty pounds of fertilizers and between five and ten pounds of pesticides every year.[8] The National Academy of Science found that homeowners use up to ten times more chemical pesticides per acre than do farmers.[9] In Connecticut, homeowners use 61 percent of the pesticides applied in the state.[10]

Once applied to a lawn, fertilizers and pesticides can follow a variety of paths and have a variety of unanticipated environmental effects. Through the lawn's connection with the air stream and flowing water, chemicals can move into and affect distant ecosystems.

Fertilizers Humans often find rapid growth desirable in the landscapes they manage; lawns are no exception. The lush green of the Industrial Lawn is a sign of high growth rates. Growth is the rate at which solar energy, captured in photosynthesis, is stored in plants and is measured as a gain in weight. In naturally occurring ecosystems, plants seldom operate for long at maximum growth rates because some factor eventually limits production. Scientists have found that if the limiting factor is added to the ecosystem, high growth rates will resume. Thus, fertilization increases growth by removing the effect of limiting factors.

Naturally occurring ecosystems usually retain most of their nutrients within the system, but human-managed systems can lose substantial amounts. Many nutrients are lost through harvesting, be that fruits, vegetables, grains, logs, or the clippings and leaves we remove from the lawn. Nutrients can also be lost through erosion, a significant problem for the American farmer. The application of synthetic fertilizers replaces lost nutrients. To ensure maximum productivity farmers and lawn owners often add more fertilizer than plants are capable of assimilating into their tissues. For example, plants will immediately use only a part of the nitrogen present in fertilizer; some will be incorporated into the soil, some may change form and be lost as a gas, and some may be lost in drainage water.[11]

The environmental impact of overfertilization can be seen within the lawn ecosystem and in connected ecosystems like streams, lakes, and estuaries as well as in the earth's atmosphere. Excess nutrients, especially nitrogen, have an array of negative effects on the ecosystem of the lawn. Too much nitrogen can increase the grass plant's vulnerability to disease, reduce its ability to withstand extreme temperatures and drought, and discourage microorganisms that are beneficial to lawn health. Some synthetic fertilizers may acidify the soil, limiting important biological and chemical processes.[12]

Fertilizer use can be linked to changes in the earth's atmosphere. When nitrogen fertilizers break down in the soil, the gas, nitrous oxide, can be released into the air. Experts have recently pointed out that nitrous oxide is a potent greenhouse gas that contributes to climate warming.[13] Nitrous oxide is also one of several gases that, upon reaching the uppermost levels of the earth's atmosphere, act to destroy the stratospheric ozone layer that protects the surface of the earth from damaging ultraviolet radiation from the sun.[14] As we will see in a later section, nutrients from fertilizers can also pollute water.

Pesticides Three groups of organisms may be thought to "threaten" the lawn: animals (such as moles and insects), weeds, and fungi. They may bring disease or change the appearance of the lawn, disrupting the smooth, even carpet of green. Since the 1950s many homeowners have waged war on these enemies with pesticides, specifically with rodenticides, insecticides, herbicides, and fungicides. In 1990, the EPA estimated that 70 million pounds of herbicides, insecticides, and fungicides were applied to residential homes and gardens.[15] Naturally occurring ecosystems protect themselves against disease and insect outbreak in many different ways. Some plants, like milkweed, produce chemicals in their leaves that make them unpalatable to organisms that might otherwise feed on them. Leaf-eating insects may be held in check by predators or disease, but in a larger sense, natural ecosystems often maintain good health through the large number of plants and animals that make up the system. If one plant or animal is weakened by insects or disease, another type of organism can take its place. Thus natural ecosystems usually do not put all their eggs in one basket.

Although a good, healthy lawn is the homeowner's goal, the encouragement of diversity in the lawn is often frowned upon. A lawn composed of grass species alone is the desired goal in the Industrial Lawn. In this instance human goals are in direct opposition to nature's management scheme.

In their book *Lawn Care,* H. F. and J. M. Decker describe potential environmental problems associated with continuous pesticide use.[16] These problems are outlined below and are illustrated in the box, Chemical Dependence:

1. Pest resistance: With the continued use of pesticides, resistant strains of target pests increase, and the pest population becomes more difficult to keep in check.

2. Inadvertent pest enhancement: When one pest is eliminated, another previously insignificant pest may attain a significant foothold.

3. Killing beneficial organisms: Many insects, soil organisms such as earthworms, bacteria, and fungi carry out functions important to the health of the lawn. There is, for example, a fungus that moves nutrients and water from the soil to the plant root. A fungicide applied to kill disease-causing fungi may also kill these beneficial fungi. As a result, the lawn owner may have to supply more water and fertilizer to achieve the desired productivity. Natural predators, such as spiders, help control harmful pests, but may become the unintended targets of pesticides. Ironically, pesticides may kill off one of the lawn manager's greatest allies, microorganisms that decompose thatch. Clearly, pesticides interfere with the lawn's natural means of maintaining good health.

4. Pesticide persistence: Until recently, many pesticides used in the United States persisted in the environment for a long time with their lethal capabilities intact. Newer pesticides have a much shorter period of potency, but they can nevertheless affect ecosystems beyond the lawn where they were applied. Pesticides may blow off site or be leached from the lawn in drainage water and end up in wells or in streams and lakes where fish and other aquatic species may be affected. Pesticides are known to kill shellfish and other species in marine environments.

PESTICIDE APPLICATION

THIS SIGN MUST REMAIN FOR 24 HOURS FOLLOWING
PESTICIDE APPLICATION

Chemical Dependence in Your Own Backyard

Perusal of any major newspaper in the United States is
sufficient to make one aware that drug dependency is a
major social problem in our country. Addiction to cocaine,
heroin, crack, and alcohol has left a trail of broken homes
and people and a society totally confused about solutions.

Yet chemical dependency is not limited to drugs; a
form of it may be present in your own backyard! The Industrial Lawn is in many respects a chemically dependent ecosystem. How does dependence develop? We could
enter the cycle of dependence at any number of points,
but let us start with the removal of grass clippings. Clippings are rich in nutrients and if left on the lawn, they
quickly decompose, making nutrients available to the
next generation of grass plants. Removal, however, neces-
sitates the use of fertilizers to replace the lost nutrients.
Fertilizers often create environmental conditions in the
soil that are inimical to decomposing organisms like
earthworms and microbes; the weakening of these organisms leads to even less decomposition and the need for

even more fertilizer. Excessive fertilization can cause the grass blades to grow at the expense of the roots, making the grass plant more susceptible to drought and necessitating the use of irrigation to maintain growth and greenness. Pesticides are applied to control insects, fungi, and weeds, but their effects are often not limited to specific pests, and many nontargeted species may be damaged, weakened, or eliminated. Such nontargeted organisms may include those that carry out decomposition, the natural predators of insects destructive to grass and other desirable plants, and the fungi that contribute to plant growth by supplying nutrients. These chemicals weaken both the ecosystem's natural defenses against insects and disease and its naturally occurring nutrient cycling; to keep the altered system going, still more pesticides and fertilizers are required. Like any heroin addict, your lawn continually needs a fix to keep up its brave green front. But it is possible to break the cycle of dependency with all of its negative implications for our biosphere: just say no to chemicals.

Some of the synthetic organic chemicals used as pesticides are fat soluble and degrade slowly. As these pesticides move through the food chain, they become more and more concentrated and may even reach toxic levels in animals of prey. DDT, a pesticide used extensively in the 1950s and 1960s, washed into aquatic environments where it was taken up by small algal plants. These were consumed by small fish which were in turn eaten by larger fish and eventually by birds of prey far removed from the original site of pesticide application. High concentrations of DDT in their fatty tissues caused birds like the osprey to produce thin-shelled eggs that cracked easily causing reproductive failure and an alarming drop in their population size. The banning of DDT in the United States has allowed some bird populations to recover, but chemicals that are hazardous to wildlife continue to be used. For example, one of the most widely used insecticides, diazinon, kills waterfowl and other bird species. The EPA banned use of diazinon on golf courses and sod farms after reports about birds dying. But, one may fairly ask, how do birds distinguish a diazinon-free putting green from a large suburban front lawn that has just been treated with diazinon? The EPA's ruling is still under review.

Human Health and the Lawn

Homeowners often watch, without concern, as workers from lawn care companies don protective suits before spraying pesticides on their lawns (figure 27). This response stems from the belief that lawn chemicals are not dangerous or that the benefits of sprays simply outweigh the risk of spraying potentially health-threatening toxins in their backyards. Unfortunately, the people who choose to dismiss health concerns for themselves and their children do so for their neighbors as well, for the environment affected by their application of pesticides does not stop at their fence or their driveway.

Although our lawns conjure up images of good health, lawn care chemicals are increasingly associated with human health problems. In 1990, the United States Senate conducted hearings on the use and regulation of lawn care chemicals. Some of the findings were startling:

1. Thirty-two out of thirty-four major lawn care pesticides have not been fully assessed for their long-term effects on human health and the environment.

Figure 27. Chemical control of weeds and pests is becoming increasingly popular among homeowners. Although chemicals serve many useful functions, most yard management procedures can be accomplished without them. Photograph by Runk/Schoenberger/Grant Heilman Photography, Inc.

2. Of the people using pesticides, 50 percent do not read the warnings on the containers that are designed to protect human health.

3. Commonly used lawn care chemicals have been implicated in a number of human health problems. Individuals testified that they experienced severe nervous system reactions, including nerve damage, after exposure to some chemicals. One man testified that his brother had died after playing golf on a course that had just been sprayed with a fungicide. There were other reports of acute toxic responses to lawn care chemicals as well as complaints of nausea, rashes, and headaches. Reports of pets dying after exposure were not uncommon.

Chemical manufacturers tend to dismiss reports of illnesses or death as isolated and unsubstantiated cases,

> but many well-documented cases of pesticide poisonings
> exist and it would seem prudent to err on the side of cau-
> tion, if for no other reason than that our lawns are the
> playgrounds of America.

Water Supplies

"We talk [water] scarcity, yet we have set [some of] our
largest cities in the deserts, and then have insisted on
surrounding ourselves with Kentucky bluegrass [see figure
28]. Our words are those of the Sahara Desert; our policies
are those of the Amazon River."
—Richard Lamm, Governor of Colorado, 1975–1987

Earlier we touched on the energy costs involved in irrigation,
in transporting water from place to place, but there is still another problem with watering lawns. We are a nation whose water
needs are rapidly rising while available supplies are shrinking
and where regional water crises are becoming increasingly frequent. Population increases in the United States have combined
with increased per capita consumption of water to generate a
water crisis. In 1985 there were 61 percent more Americans than
in 1950, yet during the same period, public water use (which
excludes agricultural and self-supplied industrial use) rose 164
percent—a rate more than twice that of our population increase.[17] Between 1920 and 1960, reservoir capacity grew about
80 percent per decade. Dam building no longer provides the easy
solution to the water problem that it once did: not only has the
number of potential dam sites decreased in number, but the
many negative effects of dam building on the environment have
also become better known.

Water tables are falling and streamflow is decreasing in many
overutilized river basins, especially in the more arid western

Figure 28. An Industrial Lawn in the semiarid climate of Salt Lake City, Utah. Photograph by Vagn Vernegaard.

states. In the Colorado and Snake rivers, as well as in many other rivers, the decreasing water flow damages aquatic ecosystems, affecting the habitat for fish and wildlife. Such depletion also results in a less secure future for human residents of depleted areas. In the semiarid Plains states, water from the vast Ogallala Aquifer, which stretches from South Dakota to northern Texas, is being extracted at a much greater rate than it is being replaced by natural processes. In the last four decades, about 120 cubic miles or 1,300 trillion gallons have been withdrawn. Water tables in the Dallas–Fort Worth area have fallen 400 feet in the last twenty-five years. In the Tucson metropolitan area, which is completely dependent on groundwater for its water supply, natural processes are only able to replace 35 percent of the water used.[18] Long droughts in California have lowered or emptied reservoirs, resulting in the periodic cutoff of irrigation water to farmers and strict rationing plans in Southern California cities.

Water supply is not only a western problem, however. Water shortages are frequent on the East Coast as well. In the New York metropolitan area there is growing concern that the increasing demand for water coupled with few remaining prime water-yielding sites may require the use of the contaminated Hudson River as a source of drinking water. The necessary treatment facilities would cost billions of dollars.[19]

A surprising amount of water for residential use goes to watering lawns. This is especially true in drier regions. Natural water balance in a lawn is determined by the amount of rainfall received and the amount lost by evaporation and runoff. In arid regions of the Southwest, the water deficit is so great that naturally occurring vegetation is dominated by desert plants that have developed a tolerance for the harsh desert conditions. Growing a lawn under such adverse conditions requires virtually constant watering. Indeed, in the West, lawn watering can account for up to 60 percent of urban water use.

Lawn watering is not limited to arid regions, however. Casual observation in southern New England, where the climate is reasonably moist in summer, suggests that irrigation is used not only to avoid the effects of extended dry periods but also to heighten productivity; thus the sprinkler may be turned on even during periods of light rain! Lawn watering has become so common that one estimate suggests that 30 percent of the water used in East Coast urban areas is for lawns.[20] Watering lawns that we have purposely designed to be thirsty is certainly a practice that needs to be re-evaluated.

Water Pollution

When contaminated with chemicals and sediments, water becomes less usable for people and may be destructive to interconnected aquatic ecosystems. Pesticides and fertilizers used on lawns can contribute to this problem. They can travel by surface runoff or seepage through the soil to drinking water wells and other public water supplies, wetlands, streams, rivers, and lakes and even to marine environments.

Several factors determine the amount and impact of fertilizers and pesticides lost from lawns: texture of soil, porosity of the soil, climate, watering practices, adjacent land uses, proximity to vulnerable aquatic ecosystems, and the kind and quantity of fertilizer and pesticide.

Sandy or gravelly soils allow water and dissolved fertilizers and pesticides to move relatively easily from the soil surface into groundwater. This movement is of particular concern in areas with permeable sandy soils lying over groundwater aquifers that supply drinking water. Such conditions exist in Cape Cod, Massachusetts, other parts of southern New England, and in Long Island, New York. Millions living in these areas are dependent on the aquifers beneath them for a supply of clean drinking water. Concentrations of nitrate in groundwater on Long Island have increased dramatically in the last thirty years. A significant proportion of this is believed to come from lawn and garden fertilization; estimates are that 60 percent of the nitrogen applied ends up in the groundwater.[21]

Chemical contamination is a notable problem in drinking water wells. Nitrate, a form of nitrogen, is the most common contaminant. EPA surveys of groundwater wells used for drinking water in the United States indicate that 1.2 percent of community water-system wells and 2.4 percent of rural domestic wells nationwide contain concentrations of nitrate that exceed public health standards for drinking water.[22] High concentrations of nitrate in drinking water may cause birth defects, cancer, nervous system impairments, and "blue baby syndrome" in which the oxygen content in the infant's blood falls to dangerous levels.[23]

Nitrate contamination may result from a number of land uses including agricultural fertilization with nitrogen fertilizers, residential septic tanks, animal wastes like those from cattle feed lots, and from lawn fertilization.[24] Once the contaminant reaches groundwater and is detected in a well, it is often difficult to determine its source. In many areas of our country, agricultural practices are responsible for most of the nitrate that reaches groundwater and wells.[25]

Different types of nitrogen fertilizers decompose to release water-soluble nitrogen at different rates. If soluble nitrogen is released at rates faster than grass plants can take it up, the excess may find its way into the groundwater. In general, organic nitrogen fertilizers such as urea release nutrients more slowly than inorganic fertilizers such as ammonia and nitrate.

Adjacent land uses also influence the relative importance of fertilizer losses from a lawn. In residential areas with many septic tanks or in areas with intensive agriculture or other sources of nitrate pollution, lawn fertilization may account for a relatively minor part of a groundwater pollution problem. On the other hand, by reducing or eliminating the use of lawn fertilizer, the homeowner can minimize the lawn's contribution to nitrate pollution as well as save some money.

When fertilizer reaches interconnected bodies of water like streams, lakes, and estuaries damage can result. The addition of certain nutrients like nitrogen and phosphorus to bodies of water can result in excessive growth of water plants in a process called eutrophication. Initially aquatic plant life flourishes because of fertilization. When these abundant plants die and sink to the bottom, they decompose, using up the oxygen in the water. Fish and other aquatic animals need oxygen, and under the oxygen-poor conditions resulting from eutrophication, many die. Ironically, the end result of too much fertilization can be a smelly body of water that is deprived of oxygen and of many forms of life.

The Mississippi River drains the heartland of the United States and like most large rivers has experienced massive changes, including eutrophication, during the last century.[26] Nitrate concentrations in the lower Mississippi have doubled in the last forty years, and agricultural fertilization has played a large role in this. A recent study estimates that 44 percent and 28 percent, respectively, of the nitrogen and phosphorus fertilizer applied in the Mississippi watershed ends in the Gulf of Mexico. Predictably, waters low in dissolved oxygen are found in the bottom ocean waters at the mouth of the Mississippi River. Although the Mississippi example does not implicate the lawn, it does emphasize the connection between fertilization and eutrophication. Many scientists believe that a significant reduction in the eutrophication of water bodies in the United States "is not likely to occur without a reduction in fertilizer use."[27]

Pesticide contamination of groundwater is less well documented than fertilizer pollution but is of growing concern. Detectable levels of pesticides or pesticide breakdown products have been found in 10 percent of the wells in community water systems. Of these, 0.8 percent have one or more pesticides above health advisory levels. DCPA, an herbicide extensively used on home lawns, is the most commonly detected pesticide found in EPA well-water surveys.[28] Because of their chemical character and their ability to travel through the soil, today's pesticides are more likely to contaminate groundwater sources than older pesticides, which leached through the soil more slowly.[29] For example, the EPA considers the commonly used herbicide 2,4D to be a "priority leacher" that travels quickly to groundwater.[30] A component of "agent orange," a defoliant used in the Vietnam war, 2,4D has been linked to cancer and birth defects.

Although many may argue the significance of this pollution, most people agree that chemical pollution of water supplies and water bodies should be avoided. Some proportion of the fertilizers and pesticides used on lawns gets into water supplies and water bodies through runoff or leaching into groundwater. Whatever the percentages may be, it makes sense to do whatever we can to minimize this pollution. Having the greenest lawn on the block is certainly not worth contaminating one's own drinking water.

Solid Waste

We generate more than a half-ton of solid waste per person, a half-ton for every man, woman, and child, for every elderly or infirm person, 160 million tons of municipal solid waste each year. Most of this astonishing quantity of waste ends in landfills. Yard waste is the second-largest component of the waste stream; three-quarters of yard waste is grass clippings from our lawns.[31] Clippings generally get bagged in large polyurethane bags and find their way to the curb on garbage day (figure 29). Landfilling was originally viewed as an environmentally sound alternative to burning, but many landfills are now filled beyond the designed capacity. It is estimated that within twenty years 80 percent of legal landfills will run out of room.[32]

Figure 29. In the course of a year, the average American lawn owner throws away between ten and fifteen cubic yards (about three dump truck loads) of grass clippings. © Photograph by Karen Bussolini, 1992.

There is no reason for grass clippings to be considered waste. Clippings can be left on the lawn where microorganisms will decompose them, releasing nutrients that have been stored in the cut grass so that they can be used again.

Grass clippings are a source of nitrogen and other nutrients. Removal of clippings may result in a loss of up to 100 pounds of nitrogen per acre of lawn per year.[33] For the state of Connecticut, clippings removed from just 20 percent of the lawns would total 2,000 tons of nitrogen moved from lawns to landfills. To replace this loss would require about 20,000 tons of 10–10–10 fertilizer![34] An additional problem is that decomposing lawn wastes in landfills can produce methane, one of the most powerful greenhouse gases.

Other wastes associated with lawn care constitute a serious problem, too. Empty or partially empty containers of insecticides and herbicides are hazardous wastes.[35] Pesticide-laden grass clippings are another significant addition to the hazardous waste stream. In addition to constituting a danger in their own right, these toxins may interfere with the process of biological decomposition that would otherwise help to break down our garbage, thereby adding to the rapidity with which we fill up our solid waste dumps.

Species Diversity

The coming and going of species has been a normal development throughout the history of the earth. Species extinctions are caused by large catastrophes or by more gradual changes in a species' environment. For example, many believe that the dinosaurs became extinct after a meteor collided with the earth. Dust and debris from the collision clouded the entire surface of the earth, reducing the amount of sunlight available to plants. Photosynthesis decreased, reducing the amount of available food for all organisms, and the temperature at the earth's surface declined precipitously. Fossil records indicate that "natural" extinction occurs at the rate of about one species per thousand years. In turn, new species appear at a similar, slow rate.

This balance between species extinction and species creation has recently been shattered by human activity. Most scientists agree that because of human activities such as habitat destruction and pollution, the extinction rate now greatly outstrips speciation rates. Although our estimates are crude at best, even conservative estimates of current species loss are startling. In 1991, scientists Paul Ehrlich and Edward Wilson estimated that "one quarter or more of the species of organisms on Earth could be eliminated within 50 years."[36] This is mass extinction, a biological holocaust of vast proportion, and a matter of extraordinary concern to biologists everywhere.

Citizens from every walk of life have expressed grave concerns about species extinction on our planet. Discussion often focuses on tropical rain forests and how their destruction feeds the high rates of extinction. While it is true that tropical forests are among the most species-rich and most-threatened ecosystems, extinctions are not unique to the tropics and the events that lead to extinction are not always as obvious as the burning of forests. Indeed as concerned citizens who wish to encourage species diversity, we may be ignoring the best place to begin this endeavor—our own backyards.

As we create and manipulate our lawns, we affect the organisms living within them. A new housing development frequently involves the wholesale destruction of a forest, an abandoned agricultural field, or a desert. The lawn, particularly the Industrial Lawn, is a highly simplified ecosystem compared to the forest or field, which typically contain a diverse mosaic of plants and animals. Thus, the creation of a lawn is synonymous with a reduction in species diversity (figure 30).

Animal populations are often determined by the vegetation available for food and shelter. The lawn provides little in the way of food for many species. We may see raccoons, squirrels, and other small mammals traveling across our lawns, but to survive they need areas more protected than lawns for denning sites.[37]

Birds are perhaps the most entertaining and welcome forms of wildlife that inhabit and visit residential areas. Changes in bird populations provide one of the best examples of the effect of converting natural ecosystems to suburban housing and lawn ecosystems. Many studies have shown that the bird population is greater in developed, suburban areas than in undeveloped or "natural" areas.[38] For example, in a residential area in Tucson, Arizona, bird densities were observed to be twenty-six times that of surrounding desert covered with creosote bush.[39] Yet as anyone who has marveled at the numbers of seagulls at the local dump or pigeons in a park can tell you, numbers of individual birds and numbers of bird species are two different things.

The flat, uncomplicated structure of the lawn attracts ground-feeding bird species that eat seeds and insects. House sparrows,

Figure 30. Across the country, lawns have replaced native habitats like woodlands and wetlands. Animals such as this forest-dwelling wood frog cannot survive in lawn ecosystems. Photograph by Steve Zack.

starlings, rock doves, and Inca doves (in the Southwest) are among the most abundant species in urban and suburban areas.[40] Indeed, house sparrows, a species introduced from Europe, are often the first to colonize new residential areas.

The suburban lawn displaces numerous diverse habitats that characterized the site prior to development: habitats that contained a variety of nesting sites, food, and shelter from predators. The displaced habitats contained fewer birds than the developed sites, but usually many more bird species.[41]

The more native vegetation is replaced by lawn, the less habitat is available for specialist species that rely on a few native plants for habitat or foods. Around Tucson, Arizona, the desert vegetation is characterized by palo verde and saguaro cactus. As long as native vegetation remained substantial, moderate suburban development resulted in few changes in bird populations. However, when native vegetation was greatly reduced and replaced by Industrial Lawns, native birds such as the black-tailed gnatcatcher, pyrrhuloxias, brown towhees, and black-throated sparrows were entirely replaced by house sparrows and Inca doves, birds virtually absent from native vegetation. Starlings and house sparrows ventured into surrounding native vegetation and competed with native woodpeckers for nesting sites in saguaro cactus.[42] Similar results were observed in another Tucson development located in creosote bush desert. The wide-ranging loggerhead shrike and brown-headed cowbird were absent from an urban bird census where the Inca dove, house sparrow, and starling were present in abundance.[43]

It may be time to link the local extinction that occurs in our backyards with the world decline in biotic diversity. The spread of the lawn and its accompanying destruction of native habitat may have a serious cumulative effect on the nation's flora and fauna, especially on plant and animal populations of limited size that occur in small and specialized habitats. The widespread expansion of suburbia and the American lawn has ousted local bird species and allowed a few species to dominate that are particularly suited to living among humans and houses. In a comparison study in Illinois, it was noted that bird species in urban areas were constant from region to region while in all other habitats there were great regional differences. The loss of diverse vegetation in urban areas accompanied the loss of bird diversity.

If the Industrial Lawn continues to accompany the development of our nation, we can predict two results. Not only will the important environmental functions of the undeveloped lands—the control of water and nutrient cycles and of energy flow in ways beneficial to society—be greatly diminished, but more and more species of plants and animals will be restricted to smaller and smaller areas and larger and larger groups of fewer species will dominate the landscape. Already 20 million acres of residential lawns have combined with millions more acres of lawns in public parks and highway margins to displace a vast amount of native habitat for a large number and variety of plants and animals.

The Environment and the American Lawn

The American lawn has a long and noble history, and it has won a firm place in our hearts, but there is a darker side to this grassy swath. In our efforts to make it greener, to make it all grass, to keep it closely mowed, and to make it a constant companion of suburban development, we are unnecessarily contributing to some of the most severe environmental problems facing the world today.

The use of fossil fuels both directly in gasoline-powered equipment and indirectly through irrigation and the production and transport of fertilizers and pesticides contributes to regional air pollution and global warming. Excess fertilizers and pesticides wash off our lawns and run into our wells, streams, and lakes where they may contribute to major environmental and human health problems. Lawn irrigation can exacerbate already severe regional water-supply problems, and lawn wastes are major contributors to our increasingly severe national solid waste problem. Finally, the replacement of millions of acres of naturally occurring ecosystems by the American lawn during the development process plays an important role in the continuing decline of local and regional species of plants and wildlife.

This does not mean that all aspects of the lawn are negative or that the eradication of lawns would solve any of the environmental problems to which they contribute. What all this information does mean is that we are now armed with the knowledge of how best to maintain or alter our lawns with the goal of creating a more healthy, diverse environment. Knowing the ecological implications of our actions, however great or small, enables us to take action.

Chapter 5
A New American Lawn

We started this book with a history of the lawn and the story of its eighteenth-century rise to prominence as the centerpiece of a particular view of nature. We have brought the story up to the twentieth century by showing how industrialization transformed this favored piece of green in relation to the environment. Now we shall bring the lawn's aesthetic history up to the present.

The changes that brought about specialized seed varieties, fertilizers, and machines on a large scale affected not only agriculture and landscape. They also informed and transformed other areas, particularly the arts. Architecture, painting, and sculpture have all wrestled with their relationship to the industrial world and machines, which began to produce synthetic materials and chemically produced colors. Out of this struggle came the new forms that we call the modern movement. Landscape, as art, was the latest star to arrive in the constellation of the modern movement, but it lagged behind the others and saw itself more as a craft. Residential yards and gardens particularly remained as craftwork, with one exception: the lawn. The eighteenth century's favored piece of green received all the attention of industry: seed companies developed a lawn based on very few species; sod farms made the lawn a shippable, instant landscape. The lawn became, in fact, the modern landscape. And its importance went well beyond a residential application, though it has been the piece of modern landscape attached to most residential sites; it has become the favored landscape piece for all modern buildings, whether corporate headquarters or museums. The eighteenth-century lawn has taken on another meaning for the twentieth century: it has become a visual symbol of the control of living things by humans, a perfect, completely malleable piece of nature. Or so we thought.

With our growing concern about the environmental health of the planet and our emerging understanding of ecological damage that may be caused by the American lawn, some have begun to ask, can I manage or redesign my lawn in ways that minimize its negative impact on the environment and yet create an environment that meets my aesthetic and recreational needs? Or, in the broader context, can I be a good steward of that small piece of the biosphere entrusted to my care?

We will now weave together our new ecological understanding with concepts of design and management. Presenting ecological reasons and data might be considered sufficient to bring about redesign of our lawns, but ignoring landscape design principles perfected over time to fulfill ideals of beauty denies the complexity of human behavior and dooms much of ecology's hard work. If lawns cannot be replaced with landscapes of beauty and usefulness, few of us will want to change. On the other hand, landscape designers can no longer ignore the sound questioning of current landscapes and their management. Ideas about ecology do not design a landscape; ideas about beautiful landscapes as currently practiced have often had difficulty incorporating ecological ideas. Here we bring together some recommendations from ecology and art, to give all of us tools and concepts to guide us in reshaping our gardens and yards. The integration will be something that will take time and thought beyond these pages.

In the following pages we suggest several alternatives to the classic American lawn: the all-grass, free-of-pests, continuously green, frequently mowed Industrial Lawn. It is not our mission to detail every possible way of transforming a lawn into an ecologically benign and aesthetically beautiful property. Our goal is to provoke new thinking about the lawn and its connection to the larger environment, to provide alternative strategies for managing or changing our lawns with the objective of implementing a new understanding of our relationship to the world around us, and to propose a new aesthetic approach to its design. No formulas or generic solutions are offered, only brief glimpses of what might be.

Where to Go for Help

Many people, organizations, and books can provide assistance and suggestions concerning ways to change or adapt the lawn.

- Government agencies, including town conservation commissions, state cooperative extension services, or the U.S. Department of Agriculture.

- Gardening or landscape and nature sections of local newspapers, libraries, and bookstores. The notes at the end of this book provide a list of titles.

- Nongovernmental organizations that focus on different aspects of lawn care. Those cited in this book include:
  1. The Lawn Institute, P.O. Box 108, Pleasant Hill, TN 38578–0108
  2. Professional Lawn Care Association of America, 1000 Johnson Ferry Road N.E., Suite C-135, Marietta, GA 30068–2112
  3. Rachel Carson Council, Inc., 8940 Jones Mill Road, Chevy Chase, MD 20815
  4. The National Xeriscape Council, Inc., P.O. Box 163172, Austin, TX 78716–3172
  5. The Bio-Integral Resource Center, P.O. Box 7414, Berkeley, CA 94707

Alternatives cover a range of options from changing the way you care for your grass to replacing the lawn completely. However, four underlying principles unite all the alternatives we propose: (a) meet the aesthetic, environmental, and economic needs and wishes of the individual homeowner, (b) to the degree possible, shift from fossil energy to solar energy, (c) reduce the use of chemicals and irrigation water, and (d) where possible, increase biological diversity. These principles reflect a shift in priorities away from a desire for a perfectly manicured expanse of lawn toward a healthy landscape more in harmony with nature.

Considerations

For most Americans, the yard and the lawn are a part of their life as well as a considerable investment in both time and money. Before launching on a course of action, it is important to assess what you want to gain by changing the management and design of your lawn. A primary goal would be to lessen your personal contribution to deterioration of the environment, but your yard and lawn serve other personal and family needs and may play an important role in life-style and psychological health. Thus you may strive to make your yard as ecologically sound as the forest primeval, but if it does not satisfy your expectations of beauty and practicality, it is not likely to succeed. So, too, economic feasibility must be considered. Fortunately, there are management and design alternatives that simultaneously meet human desires for beauty, for a healthy environment, and may save money.

The first step in considering a change is to assess your needs. Lawns and yards serve many purposes: an outdoor canvas for gardeners, a safe playground for kids and pets, a pleasant park for entertaining friends and family. For example, you may want a lawn as a playing field for touch football or badminton, but there may be areas of little use where ground covers other than grass could be substituted.

As a start, it is a good idea to survey what you have. Figure 31 illustrates a typical Industrial Lawn design and management scheme.

Before changing your Industrial Lawn, identify how various subsections of the yard are used. Consider recreational uses, views large and small, and settings for flowers, shrubs, and vegetable gardens. Give some thought to ecological conditions: sun and shade, wet or dry spots, shallow or deep soils. By simply walking around and examining the yard, you can learn a great deal. Most important, consider the climate and think about your area's native vegetation (see figure 13). All of these steps can help you evaluate your existing lawn and landscape design and help you to design alternatives that will fit your particular situation.

Despite a desire to act independently, you may wish to consider neighborhood standards. Peer pressure and even legal regulations might affect your decisions. If you wish, you can use your more private backyard to explore very personal and dramatic changes, while making less radical but nonetheless significant changes in the front yard. You might find neighbors who will want to join you in making their yards safer for children and in increasing the populations of songbirds by reducing or eliminating the use of pesticides and by increasing the variety of shrubs and other plants attractive to birds.

Finally, there are the inevitable budget considerations. If you make extensive changes to your yard, your initial costs may be relatively high, but in the long term, you should realize savings from lower energy, fertilizer, pesticide, and water costs. Even if you choose to retain a lawn, you can immediately save money by initiating some simple management changes, such as decreasing the frequency with which you mow, water, and fertilize. For reducing your impact on the environment you will receive no tax

INDUSTRIAL LAWN

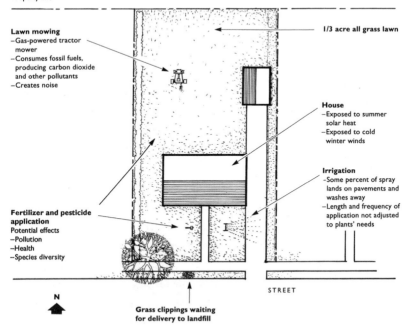

Property line

Lawn mowing
–Gas-powered tractor mower
–Consumes fossil fuels, producing carbon dioxide and other pollutants
–Creates noise

1/3 acre all grass lawn

House
–Exposed to summer solar heat
–Exposed to cold winter winds

Irrigation
-Some percent of spray lands on pavements and washes away
–Length and frequency of application not adjusted to plants' needs

Fertilizer and pesticide application
Potential effects
–Pollution
–Health
–Species diversity

STREET

N

Grass clippings waiting for delivery to landfill

Figure 31. The Industrial Lawn: a typical lot with house, front yard, backyard, and driveway. The notes highlight some of the topics raised by this book. Figures 32 and 35 depict how this yard might be changed to promote a healthier environment.

rebates or income tax credits but you will have the satisfaction of knowing that you are contributing to the creation of a richer, safer, and more diverse environment for our children.

Alternatives

Landscape designers often speak about a vision for a piece of land: patterns created by vegetation, buildings, and open space, desired uses for the property, and how these aspects fit together as a whole. Landscape design, it must be remembered, is an art. This is reminiscent of Capability Brown's view of design. He approached a landscape by looking at its capabilities: the functional and aesthetic potential of the property. What vision do you have for your yard? (1) You might want to keep the amount of lawn you have but change your management procedures. (2) Another option would be to reduce the amount of lawn and use the surrendered space for other plants or for alternative nonliving materials. (3) Finally, you might want to replace the entire lawn with other types of vegetation and landscaping materials. If you are building a new home, you have the option of preserving all or part of the existing natural vegetation.

All options involve both aesthetic and ecological dimensions. The aesthetic dimension is not formulaic. Each age has its ideal of beauty, and artists, as well as the rest of us, interpret the aesthetic ideas of an age to create an image. We point to models other than the eighteenth-century English lawn ideal and suggest you develop your own ideas or, armed with your own specific needs, preferences, and aesthetic leanings, that you work with somebody who is interested in the artistic challenge of shaping a new Freedom Lawn.

Changing Lawn Care Practice A primary goal in changing management is to make the lawn more dependent on solar energy and on the site's natural growing conditions, such as rainfall and soil, and less dependent on fossil energy and applied chemicals and water. Changes might involve selecting the right grass; reducing energy, fertilizer, pesticide, herbicide, and water inputs; or selecting an ecologically minded lawn care company. A simple management change, which would achieve the above goals, would be to switch to a Freedom Lawn. Figure 32 shows how changes in your lawn management practices can reduce the negative environmental impact.

Selecting the Right Grass If you want to keep an all-grass or nearly all-grass lawn, you might want to consider just what grasses you want. Lawn grasses include many species and many varieties within species. It is a good idea to choose varieties appropriate to your local environment and the special circumstances of your lawn. For example, you should consider local water availability. Some grasses require large amounts of water and should be avoided in areas frequently subject to drought. Some cities, Aurora, Colorado, for example, mandate the use of specific grasses to avoid the effects of drought. Aurora City Ordinance 80–47 prohibits new residences from having more than half of their landscaped area in bluegrass; any additional turf has to be planted with a drought-resistant variety with lower water requirements.[1] In other cities, such as Novato, California, homeowners are given a cash incentive for reducing the amount of grass in their yard. In a program called "Cash for Grass," homeowners receive rebates from the local water company if they replace their traditional turf lawn with drought-resistant plants. This puts cash in the homeowner's pocket and reduces the demand on an over-tapped water company.[2] Public agencies can help you choose the right type of grass for your region. For example, the Texas Water Development Board offers a guide which suggests buffalo grass for the driest parts of the state.[3]

Choosing grasses that are inappropriate to the specific conditions of your lawn's environment can lead to an unnecessary

IMPROVED MANAGEMENT

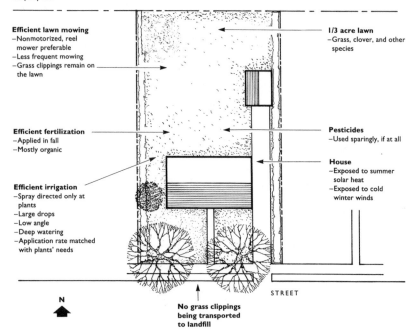

Property line

Efficient lawn mowing
–Nonmotorized, reel
 mower preferable
–Less frequent mowing
–Grass clippings remain on
 the lawn

Efficient fertilization
–Applied in fall
–Mostly organic

Efficient irrigation
–Spray directed only at
 plants
–Large drops
–Low angle
–Deep watering
–Application rate matched
 with plants' needs

1/3 acre lawn
–Grass, clover, and other
 species

Pesticides
–Used sparingly, if at all

House
–Exposed to summer
 solar heat
–Exposed to cold
 winter winds

STREET

N

**No grass clippings
being transported
to landfill**

Figure 32. Improved lawn management. The plan illustrates practices that do not change the yard's appearance, but lessen its impact on local and global ecosystems.

environmental impact. If a shade-intolerant species like Kentucky bluegrass is planted beneath a large tree, it will grow poorly. Its poor growth may cause the homeowner to add fertilizer and water, when a better solution would be to put the fertilizer and the water away and plant a grass that can tolerate some shade.

Sometimes adding a nongrass species can be advantageous. Clover can enhance a lawn in several ways. Clover has a microorganism that lives in close association with its roots. This microorganism can take nitrogen from the air and convert it to a form that can be used by the clover. When the clover dies or loses plant parts, the nitrogen is added to the soil where both grass and clover can use it again and again. This unique arrangement reduces the need for nitrogen fertilizers. Clover also has attractive white flowers, its roots bind the soil, and it can live in harmony with grass.

You might ask why this wonder plant is not a more widely accepted component of today's lawn? Several decades ago, clover was a part of most American lawns. Although some people have never appreciated its bee-attracting flowers, clover officially went out style when a large seed and chemical company launched a campaign against clover in the lawn.[4] Perhaps it is time to bring clover back.

In the Freedom Lawn, other broad-leaved plants might find a new home. Rose Marie Nichols of Nichols Garden Nursery in Albany, Oregon, and Tom Cooke of Oregon State University have developed ecology lawn mixes.[5] These mixes include grass seeds and the seeds of low-growing broad-leaved plants to provide an array of slightly different growth patterns and a lawn that tends to remain green despite variations in rainfall during the growing season.

Energy The use of fossil energy is absolutely essential to the functioning of modern society, but we now know that the by-products of combustion are also at the core of most environmental problems. By using fossil energy more carefully, conservatively, and efficiently, we can begin to reduce the environmental impact of our energy consumption.

There are a number of avenues by which we can directly and indirectly decrease the use of fossil energy associated with lawn care. First, many gasoline-powered machines are associated with lawn care: mowers, blowers, edgers, mulching machines, and so forth. We can reduce the harmful effect of these machines by using them less frequently or by not purchasing them in the first place. Electric devices are little better since the pollution is produced at the power station rather than in your backyard. As we saw in chapter 4, the tiny two-cycle engine that powers most rotary mowers is extremely polluting. Many lawns currently being cut by power mowers could be cut with a nonpowered, hand-pushed reel mower. For a lawn of modest size, a hand-pushed reel mower requires only slightly more effort than a hand-pushed power mower. The relative quiet and the fresh smell of cut grass uncontaminated by exhaust are extra rewards.

A reel mower's cutting action cuts the grass blade straight across rather than at an angle, resulting in less cut area on the grass blade and thus in a lower rate of water loss from the cut. Hand-pushed reel mowers are not effective in very tall grass, however. Manufacturers might take note and develop hand-pushed reel mowers that can handle taller grasses. Longer blades act to shade the soil and to reduce evaporation and root stress. Longer grass usually means deeper more efficient roots that can better withstand drought and disease.

If you wish to let your lawn grow to a greater height, the gasoline-powered rotary mower may be a better option. Since rotary mowers tend to rip the blades, which increases water loss, it is a good idea to keep your rotary blade sharp. If a power mower is used, a management plan that reduces the frequency of mowing will reduce your contribution to global warming and air pollution. With a nonpowered, hand-pushed mower, the frequency of cut is not a factor contributing to pollution.

In chapter 4 we discussed some of the indirect uses of fossil energy associated with lawn care. Substantial amounts of fossil energy are associated with the production and use of fertilizers and pesticides. Minimizing their use will conserve energy and reduce pollution. The same is true of irrigation. Similarly, considerable energy is required to move water around, and, as we previously discussed, in dry climates the amount of energy used in irrigation can equal that used in mowing. In expressing your concern about the management of your piece of the biosphere and your contribution to global warming and pollution, there is no decision of greater importance than your decision about your lawn's fossil energy needs.

Fertilizers Lawn fertilizers can contribute to water pollution through runoff and seepage into groundwater as well as to air pollution generated by burning fossil fuels. You might wish to consider reducing fertilizer use or eliminating it altogether as might be done with a Freedom Lawn.

The most familiar types of fertilizers are the granular fertilizers stocked on garden store shelves with the familiar nitrogen-phosphorus-potassium formulation. Usually, these chemicals are relatively soluble and thus can be quickly dissolved by rainwater and carried into the soil.

Organic fertilizers derived from decomposing organic matter can provide for most of your lawns needs. The nutrients in these fertilizers are largely insoluble and are released over time by microbial action. Some believe that lawns treated with organic fertilizers tend to "bounce back" more quickly after droughts than do lawns fertilized with synthetic chemicals. Organic fertilizers tend to promote deeper root systems. In turn, these longer roots can reach more water, even during times of drought, and thus reduce water stress on the grass plants.[6]

You can also create your own fertilizer by leaving your clippings on the lawn. They will act just like the organic fertilizers discussed above by breaking down slowly and releasing their contained nutrients. Another source of nutrients for your lawn or garden is compost. All the leaves you rake in the fall or clippings from your shrubbery or any vegetable debris, including vegetable kitchen waste, can be placed in a compost pile in the backyard. With time and an occasional turning, a rich black humus is produced. This can be applied to the lawn, shrubs, flower beds, or vegetable garden. It is among the best fertilizers available. By leaving clippings on the lawn and composting, not only are you recycling and saving on commercial fertilizer use, but you are simultaneously helping to reduce your town's solid waste problem.

Pesticides and Herbicides   Pesticides constitute a large array of chemicals designed to kill organisms that damage cultivated plants. Many also have the potential to contaminate food chains, kill nontargeted organisms, and cause human health problems. In general it is best to avoid pesticide use. If they must be used, they should be applied with extreme caution and careful forethought. Read all labels carefully and seek advice from your county agricultural agent. Your friendly hardware clerk may not be the best source of information. Before you seek advice and service from a lawn care company, find out if they have a reputation for environmental soundness.

Herbicides are pesticides designed to kill weeds, but they can also adversely affect other organisms such as beneficial soil fungi. Herbicides are often mixed with fertilizers. Weeds are plants that someone or some corporation has decided are undesirable. As we have seen, some plants considered weeds in the Industrial Lawn, such as clover, are a welcome, integral component of the Freedom Lawn. Thus your need for herbicides is strongly influenced by what you conceive to be a weed and the degree to which you wish to eradicate weeds. If you favor the

Freedom Lawn or the clover-grass lawn and are willing to tolerate a few "weeds" you will probably not need herbicides.

Be sure that any pest problem you have really needs attention;
many do not. For example, the appearance of a few insects does
not require a cascade of preventive spray.

If your problem does require attention, several categories of
treatment are available: cultural, biological, and, as a last resort,
chemical. Cultural controls involve changing procedures or plant
cover. For example, overwatering can lead to plant diseases
caused by the growth of fungi. Altering your watering regime is
easily accomplished and safer than spraying with fungicides.
Plants ill adapted to a site are sometimes attacked by pests. In
shady locations replacing sun-requiring grass with a shade-
tolerant species might solve the problem. Under heavily shading
trees, replacing all ground cover with thick mulch might avoid
disease problems.

Biological controls take advantage of natural enemies, predators, or diseases that afflict a particular pest. One of the best
known biological controls is milky spore disease, a bacterium
that attacks a wide variety of insect pests.[7] Other examples include predaceous nematodes, which can be introduced into the
soil to destroy grubs, and ladybugs, which love to eat aphids.

The combined use of cultural, biological, and chemical strategies is called Integrated Pest Management. Cultural controls are
used as a first line of defense against pests and diseases that may
be threatening the health of your lawn. These methods are safe
and generally easy. Biological controls offer a second line of defense; combined with cultural practices, these controls can solve
most of your problems. As a last resort, chemicals may be applied
to bring a pest population under control. Once the problem is in
check, cultural and biological methods can take over, eliminating
the need for continued chemical use (figure 33).

Water Water, like fossil energy, is essential to modern society.
Without an abundance of water, our society would come to a

Monitoring

–How: place coffee cans with both ends open, a few inches in the ground, fill with water, if the chinch bugs are in your soil, they will float to the top in about 5–10 minutes.

–When: once a month starting in early spring

–Where: in 4–5 random locations in your lawn

Biological controls

–Once the chinch bugs are at a minimum population level, use insect predators, available to homeowners, to maintain control.

Physical control

–Soap solution and white flannel cloth: spray infected areas with a soapy solution and cover with a flannel cloth; the bugs will crawl out of the ground and stick to the cloth.

–Dispose of the bugs so they do not reinfect the area

Cultural control

–Use slow-release nitrogen fertilizer to maintain correct levels

–Aerate soil (soil compaction results in parched lower layers)

–Plant grass varieties that are resistant to chinch bug attack

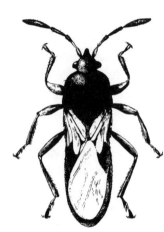

Figure 33. The treatment of chinch bugs as an example of integrated pest management (IPM). IPM uses a variety of controls applied at critical times in the pest's life cycle to keep its population to a nondamaging level. Control is achieved with a substantial reduction in pesticide use. Photograph courtesy of U.S. Department of Agriculture. For further information, see S. Daar "Safe Ways to Control Chinch Bugs on Lawns," *Common Sense Pest Quarterly* 2, no. 1 (1986): 20–22.

screeching halt. The water we use comes from precipitation. Most water that we use is fairly recent precipitation that flows in streams and rivers or is stored in reservoirs or rapidly recharged groundwater aquifers. Some water comes from deep aquifers and is a mixture of fairly recent water and older water that entered the ground centuries ago, like the water in the Ogallala Aquifer in the central United States. If the current removal rate exceeds the current recharge rate, the aquifer is on its way to exhaustion.

We can think in terms of how much runoff from the land is needed to meet the per capita use of water. Runoff is the amount of water left after water used by plants and by evaporation is subtracted from precipitation, and per capita use includes not only water used for personal uses but also the water used by industry, electrical energy production, commercial cooling, agriculture, lawn irrigation, fire fighting, and street cleaning. To meet the per capita water needs of a person living in an arid region with relatively low rainfall and relatively little runoff, a vast land area is needed. For example, the water needs of Los Angeles are drawn from as far away as northern California and Colorado. New York City, on the other hand, because of its greater rainfall draws its water supply from about the eastern third of the state. Despite the abundance of rainfall in New York, New York City, like Los Angeles, is deeply concerned about its water supplies. In March 1992, the reservoirs that supply New York City and that usually refill during the winter months were far below expected levels, raising the potential of serious shortages in the coming summer.

The dams and reservoirs built to meet the water needs of our extensive urban areas represent some of the great engineering accomplishments of our age, but the potential for increasing urban water supplies without further damaging aquatic environments is not great. Since we cannot easily or cheaply increase our environmentally sound water supplies, the best course of action is to conserve water by using it more efficiently and by minimizing nonessential uses. Throughout the country we see many evidences of this policy of water conservation.

Lawn irrigation can be considered a nonessential use, and in humid regions many lawns flourish without any irrigation. As mentioned earlier, saving water may be a matter of selecting drought-resistant grass varieties, or tolerating some brown in your lawn knowing that once rainfall returns the lawn will green up. Sometimes irrigation may be needed; for example, when a new lawn is being established, the soil should be kept moist. In making the transition away from an Industrial Lawn, some may wish to continue irrigating but may choose to do so more efficiently and more modestly. This requires following a few simple rules. • Apply only as much water as needed. Studies have found that many homeowners apply twice as much water as lawns need.[8] Residents participating in a study in Logan, Utah, watered their lawns with little regard to the actual requirements of the grass.[9] • Know how much water your sprinkling system delivers per hour. • Have a good idea of how fast water will sink into the soil and do not apply water at a faster rate. • Water in the evening. • Water less often but deeply rather than more often but shallowly. Figure 34 depicts root growth depths with different watering regimes. Much of this can be learned by personal experimentation, but you may wish to seek the advice of a lawn care professional known to be interested in water conservation.

Another strategy for reducing water use in your yard calls for clustering plants with the same water demands in the same place or hydrozone. This method, called hydrozoning, allows you to more easily match water applications with the water requirements of plants.[10]

Another conservation method involves using so-called grey water from showers, washing machines, and sinks in your yard as long as it not prohibited by law. In one home in Pebble Beach,

Plate 5. A Mediterranean courtyard. This southern European residential landscape has been adapted with great success to southern California, which also has a Mediterranean climate, and with moderate success to the desert Southwest, which is drier. Reproduced with permission from *Earthly Paradise: Garden and Courtyard in Islam* (Berkeley and Los Angeles: University of California Press, 1980). © Photograph by Jonas Lehrman.

Plate 6. In the Southwest, adobe buildings and earthen courtyards with local desert plants were responses to the ideas of the Arts and Crafts Movement of the early 1900s. Later, water scarcity and legislation restricting water use in several municipalities reinforced the earlier movement, producing a landscape with an aesthetic differing markedly from that of the lawn. Homeowners here landscape their yards with drought-resistant desert plants, crushed stone, and sands of different colors. These landscape designs make little or no demands on local water supplies. © Photograph by Karen Bussolini, 1992.

Plate 7. Colonial New England gardens were often formally laid out in planting beds of herbs, vegetables, and flowers with the beds surrounded by fieldstone or plank walks. The entire garden was enclosed by a wooden fence. Today, there are many modern restorations of this traditional form that offer an alternative to the lawn. Thomas Hyatt House, Ridgefield, Connecticut, designed by Patricia O'Donnell, ASLA. © Photograph by Diane Nowotarski Hobé.

Plate 8. In Milwaukee, Wisconsin, Hattie Purtell has planted native prairie grasses and flowers. Mrs. Purtell worked with landscape designer Judith Stark to create a landscape that used plants native to the area. Photograph by Tony Casper.

Plate 9. In areas with abundant rainfall, such as the Pacific Northwest, moss can be an alternative to grass. This photo of an area of irish moss was taken at the Bloedel Reserve, Bainbridge Island, Washington, and was designed by Richard Haag, FASLA.
© 1992 Mary Randlett.

Plate 10. In the Southeast, homeowners may choose to keep their yards wooded, thereby eliminating the need for mowing, applying fertilizers and pesticides, and irrigating. Photograph by Frank Golley.

light watering

thorough watering

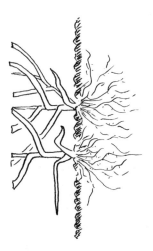
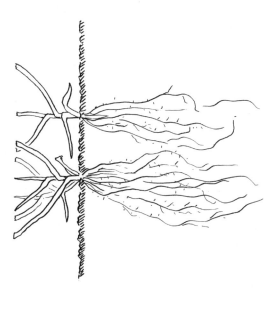

Figure 34. Your watering regime can affect a grass plant's root system and its response to drought. Frequent light watering will produce shallow roots, making the plant very susceptible to drought. Less frequent but deep watering produces deep roots that are able to withstand modest drought. © Drawing by Lauren Brown.

California, a gravity system uses a 500-gallon tank to collect water from the roof and shower for use in the garden.

However you do it, reducing the amount of water used to irrigate lawns can make a noticeable contribution to the conservation of water supplies and, indirectly, to a reduction in our use of fossil energy.

Choosing a Lawn Care Company Although many people enjoy yard work and look forward to an afternoon spent outside, 9.5 million Americans would just as soon leave these tasks to someone else.[11] Today, the service provided by lawn care companies appears to be affordable and simple. Many companies have reputable lawn care practices and can provide a homeowner with a carefree healthy lawn. Other companies, however, can lead you down the road to what might be called the lawn care equivalent of drug addiction,[12] where to achieve the continuously green lawn you must repeatedly apply chemicals, which requires that you must add water, which then requires more chemicals, more water, more chemicals, more water, ad infinitum.

When hiring a lawn care company, remember that your lawn is your property, your little piece of the biosphere, and that you have the final word. Two things should be kept in mind: greenscam and overselling. Greenscam is a tactic used by unscrupulous companies who, through false or incomplete labeling, attempt to use your concern for healthy, environmentally sound products to get you to buy products that have no particular health or environmental virtue and may, in fact, have negative side effects. Overselling is the natural tendency of any salesclerk who wishes to increase sales.

Some lawn care companies have the right environmental rhetoric but follow practices that do not accord with environmental objectives. You are told one thing, but the actuality is quite different. Other companies may try to persuade you that a pesticide application is necessary when you have no pests, or they may advocate two or more fertilizer applications where none or one would do.

In choosing a lawn care company, whether it be a nationally recognized chain or a local concern, you must distinguish between businesses that will make environmentally sound decisions about your lawn and those with other objectives such as selling you an Industrial Lawn or selling some environmentally unsound practices under the guise that they are sound. The best way to make intelligent decisions about lawn care companies is to know your lawn care objectives and the management alternatives you would adopt if you were caring for your lawn yourself.

Questions to Ask When Choosing a Lawn Care Company

What chemicals do they use?

Will they provide you with a list of the chemicals?

Will they provide you with information about the chemicals, such as the organisms each chemical affects or how long each chemical persists in the environment?

Does the company spray chemicals on the whole yard, or do they specifically target problem areas?

How many hours do you have to keep off the lawn after treatment, and does the company post this information? (Many states now have laws requiring a company to post a notice on a treated lawn. Some states require lawn companies to notify neighbors, if desired, before spraying.)

Will the company alter operations according to your specific concerns?

Will they provide a written contract, specifying what is to be sprayed and when, with a cancellation clause?

Are they familiar with Integrated Pest Management practices?

The Freedom Lawn   One of the best ways of changing your management practices and reducing your environmental impact on the biosphere is to switch to a Freedom Lawn. In humid regions, if you abandon or reduce the strict demands of the Industrial Lawn for frequent fertilization, applications of various pesticides, frequent watering, and rigorous mowing, your lawn will gradually become a Freedom Lawn. Bare spots may develop, but since the lawn is under a continuous bombardment of seeds from nearby herbs, shrubs, and trees, new plants will take hold, their presence determined largely by their ability to survive below the height of the mowing blade. Not only is it likely that the number of species will increase, but also the numbers of individuals in some species will increase. This applies not only to green plants but also to animals in the soil and microorganisms whose presence will respond to the cessation of chemical applications. Changes in soil tilth and aeration might be expected as soil organisms become more active. Better drainage might result. The lawn might shift back toward the more natural state, powered primarily by solar energy, watered by rainfall, and supplied with nutrients found in the soil. It would be more disease resistant because of its diversity of species. Growth rates would fall with the cessation of fertilization and irrigation; less frequent mowing would be required. The environmental impact of your lawn on air and water quality would be sharply reduced, and biodiversity would increase.

Although the Freedom Lawn is already the lawn of choice for many homeowners, its acceptance by homeowners for whom the Industrial Lawn is still the ideal will require a major change in attitude, a serious attempt to see things differently. Dandelions and crabgrass might become things of beauty and admiration and brown spots would be evidence of natural cycles.

Reducing the Amount of Lawn in Your Yard   You might ask yourself whether you need all of your lawn. Are there areas of your yard where grass is not needed? What about steep slopes, heavily shaded areas, corners of your property, heavily traveled paths? Replacing the lawn with low maintenance herbs, shrubs, or trees reduces the need for any water or chemicals you might be adding to the lawn and for the need to mow (see figure 35). In some parts of the yard, gravel, wood, or brick patios might make better sense than a lawn. Of course if you are building a new home, you may choose to preserve some of the original vegetation and blend it with lawn.

Replacement alternatives from the plant world are nearly endless. You should, however, choose plants adapted to where you live that can maintain themselves without added supplements and are appropriate to the size of your yard. These may be plants from your native region or introduced plants that have naturalized to local conditions. Such plants are not hard to obtain; there are more than one hundred native plant nurseries and mail order houses scattered throughout the country. Many are listed in *Gardening by Mail* by Barbara Barton, and in Henry Art's six-book series, *The Wildflower Gardener's Guide*.[13]

Trees and shrubs can be used to introduce both structural variety and species diversity and to attract birds and other wildlife by providing food and shelter. Plants might be used as architectural elements to create spaces or provide privacy. Plants can also be used for glare reduction, traffic control, and sound abatement. Shrubs can act as a snow fence. Well-placed trees can modify the flow of solar energy and wind and reduce a household's cost for heating and cooling.

Many homeowners might consider using part of their lawn for vegetable or flower gardens. There can be great satisfaction and pleasure in growing your own vegetables. They taste better and you can be certain that they do not contain any chemical residues. Flowers add beauty to the yard and can be cut for bouquets to bring the beauty of the yard into the home.

Brick terraces, sculptures, fences, or a potting shed are often

REDUCING THE LAWN'S PROPORTION

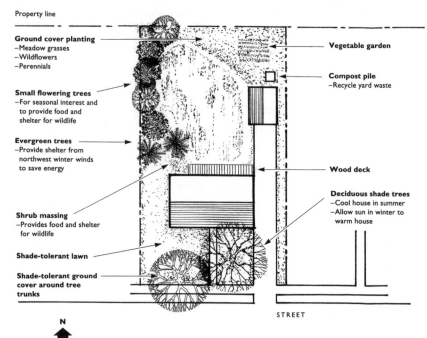

Property line

Ground cover planting
–Meadow grasses
–Wildflowers
–Perennials

Small flowering trees
–For seasonal interest and
 to provide food and
 shelter for wildlife

Evergreen trees
–Provide shelter from
 northwest winter winds
 to save energy

Shrub massing
–Provides food and shelter
 for wildlife

Shade-tolerant lawn

Shade-tolerant ground
cover around tree
trunks

Vegetable garden

Compost pile
–Recycle yard waste

Wood deck

Deciduous shade trees
–Cool house in summer
–Allow sun in winter to
 warm house

STREET

N

Figure 35. Reducing the lawn's proportion: replace parts of the lawn with shrubs, flowers, or ornamental grasses. A modest change of this kind can result in a more varied landscape as well as reduce the lawn's impact on the environment.

used as elements of the ground plan. These elements emphasize line and shape. In addition to their aesthetic value, many surfaces are both practical and colorful: bluestone, bricks, granite, gravel, and wood. They can be beneficial to the natural ecology of the yard: when properly spaced, these materials allow water to soak into the underlying soil.

All across the country people are experimenting with changing the composition of their yards and introducing native species. Start with a small area and gradually reshape the whole property. In that way, your neighbors will have time to learn what you are trying to do and may even follow suit.

Replacing the Lawn Consider taking the idea of reducing the amount of your yard planted in grass a step further: consider replacing the entire lawn. The current popularity of the lawn is closely tied in to its beauty, first forged by a handful of artists in eighteenth-century England; new concepts of yard use should not only respond to our environmental and ecological concerns but present equally powerful aesthetics. Can these alternatives be beautiful, as beautiful as our lawn? In response to this question, we present a series of images from different regions and cultures that can serve as a point of departure for thinking about the beauty of landscapes that are *not* based on the lawn.

If you wish to replace your lawn, it is important to decide what you want to achieve with your yard ecologically, aesthetically, and functionally. If you do not have the time or inclination to do the work yourself, you might choose a landscape architect, horticulturist, or artist who can give form to your new goals. Before hiring one of them you need to know if they are responsive to your ideas and issues raised by this book. Ask for examples of their work and for a description of how they would implement your new goals.

Using Ideas from Other Cultures

Landscapes of great power and beauty not based on the lawn have evolved in other times and climates. Although the Mediterranean Sea is surrounded by different cultures, the basin shares a relatively dry climate. One remarkably consistent landscape has emerged there: paved courtyards surrounded by walls and lavishly planted with vines and potted plants, where pitched roofs direct water from infrequent rains into a small pool in the courtyard center. The walls surrounding the yard provide shade and protect against drying winds. The vines planted along the walls as well as the multitude of pots fill the patio with the scent of flowers and of herbs and help to shade the walls and ground from the heat of the sun (plate 5).

The Japanese landscape tradition is derived from the Chinese, but the influence of Zen religion took the Japanese temple gardens in a much more abstract direction. Gardens, devoid of any grass or planted beds, use crushed stone or raked sand (which allow infiltration of rainwater), rocks, shrubs, trees, and a path that must be carefully followed. Groupings of shrubs and trees are skillfully layered, and beneath large trees moss can sometimes be found. The temple tradition in gardens was adapted to Japanese house gardens, which are mainly long, narrow corridors, and the careful lessons in scaling down the elements of the garden serve to create an extraordinary sensation of ample space (figure 36).

In the medieval Western tradition, gardens were located behind castle and cloister walls where protected space was at a premium. There, baskets of woven vines or stone retaining walls enclosed highly productive raised planting beds. These raised beds could be reached from both sides for planting, weeding, or cultivating and were dedicated to specific plantings: medicinal herbs, vegetables, and flowers.

Figure 36. Japanese gardens are not imitable for they are aesthetic responses to historic, cultural, and geographic conditions that are different from our own. These gardens illustrate a well-designed landscape devoid of grass and accomplished with great economy of means. We can take lessons from their minimal use of resources, their low consumption of water, their surface drainage, and their low maintenance in terms of chemicals, insecticides, and mechanical upkeep. Ko-myo-ji, Japan. Photograph by Diana Balmori.

Is there anything in these examples for our time? Perhaps nothing, if we take them literally; however, in creating new aesthetics, ones suitable to our climatic conditions and our environmental goals, universals may be extracted from each that offer wise instruction.

Today, in some of our warmer climates, much can be learned from the climatic control Mediterranean courtyards obtain with walls and trellises and plantings. The abstract crushed stone gardens of the Japanese landscape have a kinship with the environmentally sound stone and cactus yards of the Southwest. The raised planting beds of medieval origin, so easily managed and of great productivity, offer environmental advantages for small spaces or hostile environments because of their efficient use of water and nutrients; they have become the preferred form for modern urban community gardens.

In the 1920s and 1930s, across the United States, several attempts were launched at making local material the source of regional design and identity. The Arts and Crafts Movement of the late nineteenth century, which developed in England in opposition to the homogenization brought about by industrialization, stimulated interest in the regional and vernacular characteristics of place. In the Midwest this art movement took the name of Prairie School, and it produced a rich heritage of both architecture and landscape design. In California and the Southwest, it was called the Mission Style and was based on the architecture and landscape of the old Spanish missions and their courtyard gardens. Important local variants existed, especially in New Mexico. There local adobe-building traditions, both Indian and Spanish, were brought back into favor, as was planting in accordance with the desert conditions of the region. In the late 1980s with the increasing stress on water supplies, xeriscapes or desert plantings received another boost (plate 6).

A Partnership with the Region

A sense of place arises from our ability to recognize where we are in this world by the natural landforms and native species. Many areas of our country have lost their unique identity through the introduction of the Industrial Lawn and nonnative ornamental

Figure 37. Large native trees can establish a yard's character. If left undisturbed, a great variety of plants can grow on the woodland floor. In the example shown here, the backyard is considered a showcase for the diverse plant life of the Midwest. Photograph by Tony Casper.

Figure 38. Many city dwellers have uprooted most or all of their lawns to plant shrubs, flowers, and trees. Photograph by Diana Balmori.

species. The reintroduction of native plants can help re-create a sense of place reminiscent of our predeveloped landscape. Removing the lawn and some of the nonnative plants in the yard and replacing them with native plants and native vegetation patterns is one way of reestablishing a local sense of place.

Like Capability Brown's vision, which was so well suited to the English environment, but when applied elsewhere, was so often environmentally unsound, there is no single environmentally sound design that can be applied everywhere in the United States—our country is just too large and too climatically and vegetationally diverse for any single solution to work everywhere. Each region has an array of native species and natural vegetation (see figure 13) that is able to grow, survive, and reproduce using the solar energy, rainfall, and soils that the place has to offer. In our search for landscape paradigms that will both

Figure 39. Rather than maintaining an Industrial Lawn, Sue Heilman and
Walter Silver of Cambridge, Massachusetts, chose to completely replace their
lawn with a combination of tile, gravel, and border plantings to create an
intimate and easily maintained backyard space. Photograph by
Walter S. Silver.

meet our environmental needs and satisfy our search for new
aesthetics, the best advice may be to look around us.

Rather than embark on a lengthy discussion of these new locally based aesthetics, we have chosen a few images to depict
these ideas. Plates 7 through 10 and figures 37, 38, and 39 depict
nongrass landscapes from various regions of the country, each
of which is an aesthetic adaptation using native plants and
habitats.

We hope this book will inspire many Americans to reexamine
their use of that piece of the biosphere entrusted to their care. By
choosing aesthetically pleasing but environmentally sound alternatives to the classic American lawn, we can unite our environmental concerns with direct personal action. Indeed, we can

draw the line on environmental degradation in our own yard. Understanding where the lawn's popularity comes from, how the lawn fits into the global environment, and finally, what changes we can make to alter its effects gives each of us the power to improve our piece of nature. We need not cease to love the lawn. By understanding how it does and does not work, we can adapt it to our time.

Notes

Chapter 1

1. W. Whitman, *Leaves of Grass,* 1855, in *Walt Whitman: Complete Poetry and Collected Prose* (New York: Viking Press, 1982), 31.

2. J. Falk, telephone interview conducted by B. Milton, Feb. 1992.

3. T. Hiss, *The Experience of Place* (New York: Knopf, 1990).

4. W. C. Ford et al., eds., *Journals of the Continental Congress 1774–1789,* vol. 30 (Washington, D.C.: U.S. Government Printing Office, 1904–37), 230–31.

5. J. Adams, Letter to John Sullivan, May 27, 1776, in *Papers of John Adams,* vol. 4, *February–August 1776,* ed. R. J. Taylor (Cambridge: Harvard University Press, 1979), 210.

6. Weyerhaeuser Company, *The Value of Landscaping,* vol. 4 of *Ideas for Today* (Tacoma, Wash.: Weyerhaeuser Nursery Products Division, 1986).

7. For a general reference on medieval gardens, see J. Harvey, *Mediaeval Gardens* (Portland: Timber Press, 1981).

8. General references on French gardens include E. B. MacDougall and F. H. Hazlehurst, eds., *The French Formal Garden* (Washington, D.C.: Dumbarton Oaks Research Library and Collection/Trustees for Harvard University, 1974) and W. H. Adams, *The French Garden: 1500–1800* (New York: Braziller, 1979).

9. For a general reference on English gardens, see D. Watkin, *The English Vision: The Picturesque in Architecture, Landscape and Garden Design* (New York: Harper and Row, 1982).

10. S. Pugh, *Garden-Nature-Language* (Manchester, Eng.: Manchester University Press, 1988).

11. For more information on William Kent, see M. I. Wilson, *William Kent: Architect, Designer, Painter, Gardener, 1685–1748* (London:

Routledge and Kegan Paul, 1984), and J. D. Hunt, *William Kent: Landscape Garden Designer: An Assessment and Catalogue of His Designs* (London: A. Zwemmer, 1987).

12. H. Walpole, *The History of the Modern Taste in Gardening/Journals of Visits to Country Seats* (New York: Garland, 1982), 264.

13. G. B. Tobey, Jr., *A History of Landscape Architecture: The Relationship of People to Environment* (New York: American Elsevier Publishing Co., 1973).

14. David Watkin, *The English Vision* (New York: Harper & Row, 1982), 181–82.

15. R. Turner, *Capability Brown and the Eighteenth-Century English Landscape* (New York: Rizzoli, 1985).

16. Y. Abrioux, *Ian Hamilton Finlay: A Visual Primer* (Edinburgh: Reaktion Books, 1985), 38.

17. W. Cronon, *Changes in the Land: Indians, Colonists and the Ecology of New England* (New York: Hill and Wang, 1983).

18. T. Jefferson, *A Tour to Some of the Gardens of England/Travel Journals* (reprint from *The Writings of Thomas Jefferson* [Washington, D.C.: U.S. Department of State, 1853–54, vol. 9] in *Thomas Jefferson: Writings* [New York: Library of America, 1984]), 627.

19. American Society of Landscape Architects, *Colonial Gardens: The Landscape Architecture of George Washington's Time* (Washington, D.C.: George Washington Bicentennial Commission, 1932).

20. K. Jackson, *The Crabgrass Frontier: The Suburbanization of the United States* (New York: Oxford University Press, 1985).

21. J. B. Jackson, "The Public Landscape," in *Landscapes: Selected Writings of J. B. Jackson,* ed. E. H. Zube (Amherst: University of Massachusetts Press, 1970), 157–58.

22. Jackson, *Crabgrass Frontier,* 60.

23. J. Loudon, *Ladies' Companion to the Flower-Garden* (London: Bradbury & Evans, 1841), 207.

24. S. O. Beeton, *Beeton's Dictionary of Everyday Gardening* (London: Ward, Lock and Co., 1874) cited in C. Thacker, *History of Gardens* (Berkeley: University of California Press, 1979), 233.

25. A. J. Downing, *A Treatise on the Theory and Practice of Landscape Gardening, Adapted to North America* (London: George Putnam, 1841; Washington, D.C.: Dumbarton Oaks Research Library and Collection/Trustees for Harvard University, 1991).

26. C. K. Doell, *Gardens of the Gilded Age: Nineteenth-Century Gardens and Homegrounds of New York State* (New York: Syracuse University Press, 1986), 6.

27. Jackson, *Crabgrass Frontier,* 58.

28. Jackson, *Crabgrass Frontier,* 59.

29. Jackson, *Crabgrass Frontier,* 59.

30. B. Kelly, "Art of the Olmsted Landscape," in *Art of the Olmsted Landscape,* ed. B. Kelly, G. T. Guillet, and M. E. W. Hern (New York: New York City Landmarks Preservation Commission and the Arts Publisher, 1981).

31. For a general reference on A. J. Davis, see Jackson, *Crabgrass Frontier,* and J. Kastner, "Alexander Jackson Davis," in *Three Centuries of Notable American Architects*, ed. J. J. Thorndike, Jr. (New York: American Heritage Publishing Co., 1981).

32. R. Cheatle, P. Farrant, and I. Latham, "Riverside, 1869," in *The Anglo-American Suburb,* ed. R. A. M. Stern and J. M. Massengale (New York: St. Martin's Press, 1981), 24.

33. W. Whyte, *City: Rediscovering the Center* (New York: Doubleday, 1988), 123.

Chapter 2

1. "Weed All About It," *People Weekly* 30 (1988): 45; Walter Stewart, telephone interview conducted by J. H. Connolly, Feb. 1991.

2. M. Pollan, "Why Mow?" in *Second Nature,* Michael Pollan (New York: Atlantic Monthly Press, 1991), 63–64.

3. Story and quotations taken from letter received from Murray Blum, dated Apr. 23, 1992.

4. Joel Meisel, telephone interview conducted by J. H. Connolly, Feb. 1991.

5. C. Darwin, *The Origin of Species* (London: W. Clowes and Sons, 1859).

6. H. D. Thoreau, "Walking," in *Excursions* (Boston: Ticknor and Fields, 1863), 174, 188–90.

7. G. P. Marsh, *Man and Nature; or Physical Geography as Modified by Human Action* (New York: Charles Scribner, 1864), 36.

8. W. Vogt, *Road to Survival* (New York: William Sloane Associates, 1948).

9. A. Leopold, *A Sand County Almanac* (New York: Oxford University Press, 1949).

10. As cited in R. C. Paehlke, *Environmentalism and the Future of Progressive Politics* (New Haven: Yale University Press, 1989), 18.

11. R. Carson, *Silent Spring* (Cambridge, Mass.: Riverside Press, 1962).

12. Carson, *Silent Spring,* 27.

13. Carson, *Silent Spring,* 173.

14. Carson, *Silent Spring,* 176.

15. B. Commoner, *The Closing Circle—Nature, Man, and Technology* (New York: Alfred A. Knopf, 1971), and P. Ehrlich, *The Population Bomb: Population Control or Race to Oblivion* (New York: Ballantine Books, 1968).

16. Worldwatch Institute, *State of the World* (New York: W. W. Norton and Co., 1984–92).

17. L. Brown, "The New World Order," in *State of the World 1991,* ed. L. Brown et al. (New York: W. W. Norton and Co., 1989), 5.

18. R. Dubois, *Celebrations of Life* (New York: McGraw-Hill, 1981), 81.

Chapter 3

1. E. C. Roberts and B. C. Roberts, *Lawn and Sports Turf Benefits* (Pleasant Hill, Tenn.: Lawn Institute, 1988), 5.

2. M. Pollan, "Why Mow?" *Second Nature,* Michael Pollan (New York: Atlantic Monthly Press, 1991), 55–56.

3. Roberts and Roberts, *Lawn and Sports Turf Benefits,* 1.

4. Roberts and Roberts, *Lawn and Sports Turf Benefits,* 28.

5. A. R. Goldin, *Grass: The Everything, Everywhere Plant* (New York: Thomas Nelson, 1977), 143.

6. F. W. Ravlin and W. H. Robinson, "Audience for Residential Turf Grass Pest Management Programs," *Bulletin of the Entomological Society of America* 31, no. 3 (1985): 45–50.

7. Professional Lawn Care Association of America and the Lawn Institute, "The ABC's of Lawn and Turf Benefits" (Professional Lawn Care Association, Marietta, Ga., and Lawn Institute, Pleasant Hill, Tenn., n.d.).

8. "Sell the Environment to Increase Your Profits," *Lawn News* 7 (1991): 3.

9. National Gardening Association, *National Gardening Survey, 1991–1992* (Burlington, Vt.: National Gardening Association, 1992).

10. Goldin, *Grass.*

11. Whit Yelverton of the Fertilizer Institute, telephone interview conducted by B. Milton, Feb. 1991.

12. For statistics on the fertilizer industry, see U.S. Department of Commerce, *United States Industrial Outlook '90* (Washington, D.C.: U.S. Government Printing Office, 1990).

13. R. Ringer of Ringer Corporation, telephone interview conducted by B. Milton, Feb. 1991.

14. U.S. Department of Commerce, *United States Industrial Outlook '92* (Washington, D.C.: U.S. Government Printing Office, 1992).

15. U.S. General Accounting Office, *Lawn Care Pesticides: Risks Remain Uncertain While Prohibited Safety Claims Continue,* A Report to the Chairman, Subcommittee on Toxic Substances, Environmental Oversight, Research and Development, Committee on Environmental and Public Works, U.S. Senate, 101st Cong., 2d sess., 1990, #RCED-90-134.

16. D. Pimentel, "The Dimensions of the Pesticide Question," in *Ecology, Economics, Ethics: The Broken Circle,* ed. F. H. Bormann and S. R. Kellert (New Haven: Yale University Press, 1991), 59.

17. Cited by "The Dangers of Lawn Care," compiled by the editors of *Organic Gardening* (Emmaus, Penn.: Rodale Press, Jan. 1989), 1.

18. Xeriscape is a registered trademark of the National Xeriscape Council.

19. "Lawn Care Industry Dilemma," *American Horticulturist* 69, no. 11 (1990): 4.

20. Turfgrass Council of North Carolina, *North Carolina Turfgrass Survey* (Raleigh, N.C.: North Carolina Crop and Livestock Reporting Service, 1987), 37.

21. U.S. Department of Commerce, *United States Industrial Outlook '92*.

22. U.S. Department of Commerce, *United States Industrial Outlook '90*.

23. U.S. Department of Commerce, *United States Industrial Outlook '92*.

24. Professional Lawn Care Association, News Release (Marietta, Ga., Dec. 1, 1990), 4.

25. R. Krughoff, "Lawn Care Firms," *Washington Consumers Checkbook* 7, no. 3 (1990): 49–56.

26. "The Dangers of Lawn Care," compiled by the editors of *Organic Gardening* (Emmaus, Penn.: Rodale Press, Jan. 1989), 1.

27. U.S. General Accounting Office, *Lawn Care Pesticides,* 15.

Chapter 4

1. California Air Resources Board, *Technical Support Document for California Exhaust Emission Standards and Test Procedures for 1994 and Subsequent Model Year Utility and Lawn and Garden Equipment Engines* (El Monte: California Air Resources Boards, 1990).

2. City of Irvine, Community Development Department, *Sustainable Landscaping Guideline Manual,* 1991 Draft (Irvine, Calif.: Community Development Department, 1991).

3. F. Lyman, *The Greenhouse Trap: What We Are Doing to the Atmosphere and How We Can Slow Global Warming,* World Resources Institute Guide to the Environment (Boston: Beacon Press, 1990), ix.

4. Jerry Martin, spokesperson for the California Air Resources Board, telephone interview conducted by L. Vernegaard, Sept. 1992.

5. California Air Resources Board, *Technical Support Document.*

6. U.S. Environmental Protection Agency, *Nonroad Engine and Vehicle Emission Study,* Report No. 21A-2001 (Washington, D.C.: Environmental Protection Agency, Office of Air and Radiation, Nov. 1991).

7. M. Talbot, "Ecological Lawn Care," *Mother Earth News* 123 (1990): 60–73.

8. W. Schultz, "Natural Lawn Care," *Garbage* 2, no. 4 (1990): 26–34.

9. U.S. Senate, *The Use and Regulation of Lawn Care Chemicals: Hearing before the Subcommittee on Toxic Substances, Environmental Oversight, Research and Development of the Committee on Environment and Public Works, U.S. Senate. March 28, 1990,* Senate Hearing 101–685 (Washington, D.C.: U.S. Government Printing Office, 1990), 1.

10. C. R. Frink, *Uses of Pesticides in Connecticut,* Connecticut Agriculture Experiment Station Bulletin No. 848 (Hamden: Connecticut Agricultural Experiment Station, 1987).

11. O. P. Engelstad, ed., *Fertilizer Technology and Use,* 3d ed. (Madison, Wisc.: Soil Science Society of America, 1985).

12. W. Schultz, *The Chemical-Free Lawn* (Emmaus, Penn.: Rodale Press, 1989).

13. S. L. Tisdale, W. L. Nelson, and J. D. Beaton, *Soil Fertility and Fertilizers,* 4th ed. (New York: Macmillan, 1985). See also Lyman, *Greenhouse Trap*, and Engelstad, ed., *Fertilizer Technology and Use.*

14. Engelstad, ed., *Fertilizer Technology and Use,* and Lyman, *Greenhouse Trap.*

15. U.S. Senate, *Use and Regulation of Lawn Care Chemicals,* 104.

16. H. F. Decker and J. M. Decker, *Lawn Care: A Handbook for Professionals* (Englewood Cliffs, N.J.: Prentice Hall, 1988).

17. F. VanderLeeden, F. L. Troise, and D. K. Todd, *The Water Encyclopedia* (Chelsea, Mich.: Lewis Publishers, 1990).

18. S. Postel, "Managing Freshwater Supplies," in *State of the World,* ed. L. Brown (New York: Norton and Co., 1985).

19. P. Eaton, "How Bad Is New York's Environment?" *New York Magazine,* April 16, 1990.

20. "We Come Not to Bury the Lawn," *American Horticulturist* 69, no. 11 (1990): 2.

21. W. J. Flipse, Jr., B. G. Katz, J. B. Lindner, and R. Markel, "Sources of Nitrate in Groundwater in a Sewered Housing Development, Central Long Island, New York," *Ground Water* 22, no. 4 (1984): 418–26.

22. U.S. Environmental Protection Agency, *National Pesticide Survey: Summary Results of EPA's National Survey of Pesticides in Drinking Water Wells* (Washington D.C.: Environmental Protection Agency, Office of Water, Office of Pesticides and Toxic Substances, 1990), 3.

23. A. M. Petrovic, "Golf Course Management and Nitrates in Groundwater," *Golf Course Management,* Sept. 1989, 54–64.

24. A. J. Gold, W. R. DeRagon, W. M. Sullivan, and J. L. Lemunyon, "Nitrate-Nitrogen Losses to Groundwater from Rural and Suburban Land Uses," *Journal of Soil and Water Conservation* 45, no. 2 (1990): 305–10, and World Resources Institute, in collaboration with the United Nations Environment Programme and the United Nations Development Programme "Freshwater," in *World Resources 1992–93,* ed. A. L. Hammond (New York: Oxford University Press, 1992).

25. T. G. Morton, A. J. Gold, and W. M. Sullivan, "Influence of Overwatering and Fertilization on Nitrogen Losses from Home Lawns," *Journal of Environmental Quality* 17, no. 1 (1988): 124–30, and Gold, DeRagon, Sullivan, and Lemunyon, "Nitrate-Nitrogen Losses to Groundwater."

26. R. E. Turner and N. N Rabalais, "Changes in Mississippi River Quality This Century," *BioScience* 41, no. 3 (1991): 140–47.

27. Turner and Rabalais, "Changes in the Mississippi River Quality This Century," 144.

28. U.S. Environmental Protection Agency, *National Pesticide Survey*, 1–2.

29. R. F. Carsel and C. N. Smith, "Impact of Pesticides on Ground Water Contamination," in *Silent Spring Revisited,* ed. G. J. Marco, R. M. Hollingworth, and W. Durham (Washington, D.C.: American Chemical Society, 1987).

30. J. M. Halstead, W. R. Kearns, and P. D. Relf, "Lawn and Garden Chemicals and the Potential for Groundwater Contamination," in *Proceedings of Ground Water: Issues and Solutions in the Potomac River*

Basin/Chesapeake Bay Region (Washington, D.C.: National Water Well Association, 1989), 355–69.

31. U.S. Congress, Office of Technology Assessment, *Facing America's Trash: What Next for Municipal Solid Waste?* OTA-O–424 (Washington, D.C.: U.S. Government Printing Office, 1989), and B. Gavitt, "Recycling Clippings Eases Pressure on Landfills," *Turf* 4, no. 2 (1991): 20–22.

32. U.S. Congress, *Facing America's Trash.*

33. W. Dest, interview conducted by L. Vernegaard, Mar. 1991.

34. Gavitt, "Recycling Clippings Eases Pressure on Landfills," 20–22.

35. U.S. Congress, *Facing America's Trash.*

36. P. Ehrlich and E. O. Wilson, "Biodiversity Studies: Science and Policy," *Science* 253 (1991): 760.

37. D. L. Cauley, "Urban Habitat Requirements of Four Wildlife Species," in *Wildlife in an Urbanizing Environment,* ed. J. Noyes and D. Proqulski (Amherst: University of Massachusetts Cooperative Extension, 1974).

38. K. V. Rosenberg, S. B. Terrill, and G. H. Rosenberg, "Value of Suburban Habitats to Desert Riparian Birds," *Wilson Bulletin* 99, no. 4 (1987): 642–54.

39. J. T. Emlen, "An Urban Bird Community in Tucson, Arizona: Derivation, Structure, Regulation," *Condor* 76, no. 2 (1974): 184–97.

40. S. W. Aldrich and R. W. Coffin, "Breeding Bird Populations from Forest to Suburbia after Thirty-seven Years," *American Birds* 34, no. 1 (1980): 3–7, and J. H. Falk, "Energetics of a Suburban Lawn Ecosystem," *Ecology* 57 (1976): 141–50.

41. D. N. Jones, "Temporal Changes in the Suburban Avifauna of an Inland City," *Australian Wildlife Research* 8 (1981): 109–19, and P. Mason, "The Impact of Urban Development on Bird Communities of Three Victorian Towns—Lilydale, Coldstream and Mt. Evelyn," *Corella* 9, no. 1 (1985): 14–21.

42. R. C. Tweit and J. C. Tweit, "Urban Development Effects on the Abundance of Some Common Resident Birds of the Tucson Area of Arizona," *American Birds* 40, no. 3 (1986): 431–36.

43. Emlen, "An Urban Bird Community in Tucson, Arizona," 184–97.

Chapter 5

1. Aurora City Ordinances, Aurora, Colorado, 1992.

2. M. Amato, "Novato's Disappearing Lawns: Restating the Case for Lawnless Xeriscape," *Garbage* 2, no. 4 (1990): 34.

3. Texas Water Development Board, *A Homeowner's Guide to Water Use and Water Conservation* (Austin: Texas Water Development Board, 1990).

4. W. Schultz, *The Chemical-Free Lawn* (Emmaus, Penn.: Rodale Press, 1989).

5. "A New Look for Lawns," *American Horticulturist* 69, no. 11 (1990): 3.

6. J. Burnett, "Organic Lawn Care," *Organic Gardening* 37, no. 5 (1990): 70.

7. G. Bugbee, Connecticut Agricultural Experiment Station, interview conducted by J. Greenfeld, Feb. 1991.

8. "How Much Water Does Your Lawn Really Need?" *Sunset,* June 1987, pp. 213–19.

9. R. Aurasteh, M. Jafari, and L. S. Willardson, "Residential Lawn Irrigation Management (Homeowner Water Management, Utah)," *Transactions of the American Society of Agricultural Engineers* 27, no. 2 (1984): 470–72.

10. B. K. Ferguson, "Water Conservation Methods in Urban Landscape Irrigation: An Exploratory Overview," *Water Resources Bulletin* 23, no. 1 (1987): 147–52.

11. U.S. Department of Commerce, *United States Industrial Outlook '90.* (Washington, D.C.: U.S. Government Printing Office, 1990).

12. Professional Lawn Care Association of America, News Release (Marietta, Ga., Dec. 1, 1990), 4.

13. B. J. Barton, *Gardening by Mail: A Source Book,* 3d updated and enl. ed. (Boston: Houghton Mifflin, 1990); and Henry W. Art, *The Wildflower Gardener's Guide,* Midwest, Great Plains, and Canadian Prairies Edition. A Garden Way Publishing Book (Pownal, Vt.: Storey

Communications, 1990–91). Five separate editions exist for the Midwest, Great Plains, and Canadian Prairies; Pacific Northwest, Rocky Mountain, and Western region; California, Desert Southwest, and northern Mexico; Northeast, Mid-Atlantic, Great Lakes, and Eastern Canada; and Southeast and Gulf Coast. Also see Henry W. Art, *The Wildflower Gardener's Guide: 101 Native Species and How to Grow Them* (Pownal, Vt.: Storey Communications, 1990).

Index